Mining and African Urbanisation

Juxtaposing literatures on urbanisation and mining at a time when small-scale artisanal as well as large-scale mining operations are transforming many African economies, this book focuses on the interplay of Sub-Saharan Africa mining and urbanisation in the context of global shifts in capital and labour flows. Classically, urbanisation has been identified with industrial expansion, but mining is a distinct subset of industrial activity, involving artisanal and large-scale mining.

Case studies of a wide variety of countries with long historical experience of large-scale mining (South Africa, Ghana, Angola, Democratic Republic of Congo, Zambia and Botswana), as opposed to more recent experiences of artisanal mining (Mozambique, Tanzania, and Sierra Leone), reveal that the mining surge in some countries and the slow-down in others where mining was formerly dominant encompasses a wide range of urban outcomes. In view of the cyclical boom-and-bust nature of mining activity and the sector's dependence upon finite resources and exposure to world market fluctuations, this book probes settlement patterns and welfare dimensions of urban change associated with African mining amidst an unprecedented spiral in global mineral prices.

This book was published as a special issue of the *Journal of Contemporary African Studies*.

Deborah Fahy Bryceson is a Reader in Urban Studies at the School of Geographical and Earth Sciences, University of Glasgow, UK.

Danny MacKinnon is a Senior Research Fellow in Urban Studies at the School of Geographical and Earth Sciences, University of Glasgow, UK.

Mining and African Urbanisation
Population, Settlement and Welfare Trajectories

Edited by
Deborah Fahy Bryceson and Danny MacKinnon

Routledge
Taylor & Francis Group

LONDON AND NEW YORK

First published 2014
by Routledge
2 Park Square, Milton Park, Abingdon, Oxon, OX14 4RN

Simultaneously published in the USA and Canada
by Routledge
711 Third Avenue, New York, NY 10017

Routledge is an imprint of the Taylor & Francis Group, an informa business

This book is a reproduction of the *Journal of Contemporary African Studies*, vol. 30, issue 4. The Publisher requests to those authors who may be citing this book to state, also, the bibliographical details of the special issue on which the book was based.

British Library Cataloguing in Publication Data
A catalogue record for this book is available from the British Library

ISBN13: 978-0-415-82625-9

Typeset in Times New Roman
by Taylor & Francis Books

Publisher's Note
The publisher would like to make readers aware that the chapters in this book may be referred to as articles as they are identical to the articles published in the special issue. The publisher accepts responsibility for any inconsistencies that may have arisen in the course of preparing this volume for print.

Contents

CONTENTS

Citation Information

The chapters in this book were originally published in the *Journal of Contemporary African Studies*, volume 30, issue 4 (October 2012). When citing this material, please use the original page numbering for each article, as follows:

Chapter 1
Eureka and beyond: mining's impact on African urbanisation
Deborah Fahy Bryceson and Danny MacKinnon
Journal of Contemporary African Studies, volume 30, issue 4 (October 2011) pp. 513-538

Chapter 2
Mining, housing and welfare in South Africa and Zambia: an historical perspective
Hugh Macmillan
Journal of Contemporary African Studies, volume 30, issue 4 (October 2011) pp. 539-550

Chapter 3
The power of mining: the fall of gold and rise of Johannesburg
Philip Harrison and Tanya Zack
Journal of Contemporary African Studies, volume 30, issue 4 (October 2011) pp. 551-570

Chapter 4
Mining, welfare and urbanisation: the wavering urban character of Zambia's Copperbelt
Patience Mususa
Journal of Contemporary African Studies, volume 30, issue 4 (October 2011) pp. 571-588

Chapter 5
Of prosperity, ghost towns and havens: mining and urbanisation in Zimbabwe
Amin Y. Kamete
Journal of Contemporary African Studies, volume 30, issue 4 (October 2011) pp. 589-610

Chapter 6
Botswana's mining path to urbanisation and poverty alleviation
Thando D. Gwebu
Journal of Contemporary African Studies, volume 30, issue 4 (October 2011) pp. 611-630

Chapter 7
Unearthing treasure and trouble: mining as an impetus to urbanisation in Tanzania
Deborah Fahy Bryceson, Jesper Bosse Jønsson, Crispin Kinabo and Mike Shand
Journal of Contemporary African Studies, volume 30, issue 4 (October 2011) pp. 631-650

Chapter 8
Exploring the connections: mining and urbanisation in Ghana
Katherine V. Gough and Paul W.K. Yankson
Journal of Contemporary African Studies, volume 30, issue 4 (October 2011) pp. 651-668

Chapter 9
A tale of two cities: urban transformation in gold-centred Butembo and diamond-rich Mbuji-Mayi, Democratic Republic of the Congo
Patience Kabamba
Journal of Contemporary African Studies, volume 30, issue 4 (October 2011) pp. 669-686

Chapter 10
Angola's planned and unplanned urban growth: diamond mining towns in the Lunda Provinces
Cristina Udelsmann Rodrigues and Ana Paula Tavares
Journal of Contemporary African Studies, volume 30, issue 4 (October 2011) pp. 687-704

Chapter 11
Diamond mining, urbanisation and social transformation in Sierra Leone
Roy Maconachie
Journal of Contemporary African Studies, volume 30, issue 4 (October 2011) pp. 705-723

Eureka and beyond: mining's impact on African urbanisation

Deborah Fahy Bryceson and Danny MacKinnon

This collection brings separate literatures on mining and urbanisation together at a time when both artisanal and large-scale mining are expanding in many African economies. While much has been written about contestation over land and mineral rights, the impact of mining on settlement, notably its catalytic and fluctuating effects on migration and urban growth, has been largely ignored. African nation-states' urbanisation trends have shown considerable variation over the past half century. The current surge in 'new' mining countries and the slowdown in 'old' mining countries are generating some remarkable settlement patterns and welfare outcomes. Presently, the African continent is a laboratory of national mining experiences. This special issue on African mining and urbanisation encompasses a wide cross-section of country case studies: beginning with the historical experiences of mining in Southern Africa (South Africa, Zambia, Zimbabwe), followed by more recent mineralising trends in comparatively new mineral-producing countries (Tanzania) and an established West African gold producer (Ghana), before turning to the influence of conflict minerals (Angola, the Democratic Republic of Congo and Sierra Leone).

The upsurge of mining currently restructuring many national economies in Africa is raising important questions about priorities in natural resource utilisation and the distribution of present and future mining wealth in the countries concerned. Rarely, however, does one read or hear discussion about mining's links with urban growth and settlement patterns and the welfare and indeed rural-cum-urban lifestyle changes involved in people's migration to mining sites. This may be related to the fact that few commentators have expertise or detailed knowledge of both mining and urbanisation. So, too, in the mining sites, migration and settlement patterns have a way of changing day by day so unobtrusively that residents take little cognisance of the role they are playing in urban transformation.

The demographic and economic changes related to mining over the twentieth century have largely been ignored. Instead, Southern Africa's large-scale mining complexes dominated the literature, with their racially segregated housing and attempts to constrict African urbanisation through a bachelor wage and oscillating migration between miners' rural homes and mine sites. Less controlled, but subject to threats of violence and intimidation, the production of 'conflict minerals' associated with the financing of civil war over the last few decades has generally relied on both coerced and tribute labour, sometimes imported from neighbouring countries, while local people have fled, making their way to the safety of more densely settled urban areas. Racially segregated urbanisation is now history, while conflict mineral exploitation – with its attendant rural–urban displacement – has ended in Sierra Leone and Angola, though it persists in the Democratic Republic of Congo (DRC). By the end of the twentieth century, most African governments had adopted neo-liberal economic policies that attracted Western and Chinese mineral investment to the continent.

A new wave of urban growth and settlement experimentation is coalescing, which we refer to here as 'mineralised urbanisation'. This special collection traces various strands of the concept with the aims of: first, understanding the synergies and tensions between the processes of mining development and urbanisation; and, second, examining the current sense of 'mining place' in Sub-Saharan Africa.

The volume draws together the analytical perceptions of 18 scholars from diverse social science disciplines including geography, urban planning, demography, history, anthropology and sociology in addition to cartography and mining engineering. Most authors are nationals of the countries they are writing about. In view of the lacuna in existing literature, each contributor was asked to provide historical background, before focusing on aspects of mining's present-day impact on urban growth and welfare in their respective country case studies. As a baseline assessment of the state of play, this Introduction identifies broad historical trends and spatial patterns, providing a platform for the detailed country studies of urban mining localities and urban hierarchies that follow. The first section reviews the dynamics of mining and urbanisation elsewhere in the world. This is followed by a historical overview of African mine-led urban growth and settlement patterns. The third section reviews findings from the case studies with respect to mining's relation to urban systems, growth dynamics, a sense of urban place and the impact of mining on welfare and poverty.

Global mineralised urbanisation: dynamics and patterns

We define 'mineralised urbanisation' as the influence of mineral production cycles and commodity chains on urban growth and settlement patterns at local, regional and national level. Mineralised urbanisation relates to the force of mining booms and busts on nation-states. Mine-led population movement and investment not only play an important role in the spatial pattern of the national economy, but also influence a country's urban settlement hierarchy either directly or indirectly. Mineralised urbanisation encapsulates the changing national urban profile arising either from the 'direct growth' of mining settlements or the 'indirect growth' of other non-mining urban centres derived from the investment of mining profits (Bryceson and Mwaipopo 2010). Conversely, when mining contracts, changes in the relative distribution of population and work between urban areas or, more drastically, from urban to rural areas, will alter the urban hierarchy.

In general, the initiation of mining activities either creates new settlements in previously underdeveloped regions or enlarges existing ones (Johnston 1982). Capital is attracted by mineral resources offering high returns on investment and labour by the prospect of high wages, relative to other available employment opportunities. Mining settlements also attract various ancillary and supporting services from the supply and hiring of equipment to retail and wholesale facilities and various forms of entertainment and recreation. A key question concerns whether or not mining settlements are able in the longer term to achieve economic diversification into other industries and services, reducing their vulnerability to external demand shocks and supporting long-term sustainability. The relationship between mining and urbanisation is analysed at two primary spatial scales in this section: individual mining settlements and national urban systems. Furthermore, we endeavour to discern the impact of both large corporate mining and artisanal mining on urbanisation. The two forms of mining often exist side by side in a locality, each exerting different migration and developmental dynamics on urban settlements.

Origins and growth of mining settlements as urban localities

Mining settlements can be divided into two main categories: specialised mining settlements or alternatively service or industrial centres whose development has been shaped by trade in mining products (Barnes, Hayter and Hay 2001). The former, functioning as 'engines of capital accumulation' (Kamete, this volume), are typically characterised by rapid initial growth. They are usually new settlements located in previously rural areas, reflecting the tendency for mineral deposits to be found in areas distant from existing centres of population.

Classically this is illustrated by coal mining towns, which grew rapidly in the industrial regions of the UK and in the North Eastern and Appalachian regions of the US over the second half of the nineteenth century (Morris 1993). These colliery towns generally had the size and density of urban communities, but not their heterogeneity, lacking a range of commercial activities and an elaborate pattern of social stratification (Morris 1986).

The development of mining settlements was more rapid in the American West, where independent miners rather than companies were involved. Most famously, the California gold rush of 1849–53 spurred a massive migration of some 300,000 people from the Eastern US, Europe, Latin America and Asia (Holliday 2002). The so-called '49ers' lived in mining camps that sprang up around the goldfields, which were notorious for their poor living conditions (Brands 2002).

More recently, the American West experienced a resource boom in the late 1970s and early 1980s. The effects of the rapid economic and demographic changes taking place in mining towns at this time prompted the diagnosis of 'Gillette syndrome', following the experience of Gillette, Wyoming. This malady referred to processes of social and psychological dislocation, manifested in high rates of crime, drug and alcohol abuse, material break down, mental issues and a reduced sense of community (Kohrs 1974). Over time, however, this 'social disruption thesis' has been subjected to increased questioning and criticism on the basis of weak empirical evidence and over-generalisation from single industry case studies (Lawrie, Tonts and Plummer 2011).

Two key themes emerge from subsequent research: the need to take account of the mining resource cycle or 'resource roller coaster'; and the importance of local circumstances (Wilson 2004). One longitudinal study of four mining communities in Utah and Wyoming that had experienced rapid growth in the late 1970s and early 1980s mining boom revealed social disruption to be place- and time-contingent, with a follow-up survey in 1995 identifying improvements in social integration, trust and community satisfaction, alongside reductions in the fear of crime (Smith, Krannich and Hunter 2001).

Specialised mining towns can be distinguished from larger cities in which mining has been an important source of urban growth. Prominent examples of the latter include San Francisco and Melbourne. San Francisco served as the main supply base for the Californian gold rush, transforming it from a remote outpost of a few hundred people in 1848 to a burgeoning town of 50,000 by 1853 (Brands 2002). It grew into a major metropolis as resource extraction and commerce generated large concentrations of capital, becoming the second largest financial centre in the US by the early twentieth century (Walker 2001). In contrast to specialised mining towns, metropolitan centres like San Francisco developed a greater financial and entrepreneurial capacity, which enabled them to grow as vital nodes for the coordination and control of successive waves of capital accumulation.

Global illustrations of mine-influenced urban systems

Here, the Australian experience is of particular interest, given its heavy economic reliance on mineral wealth and vast semi-arid terrain. The mineral boom began in the 1840s with the discovery of silver and copper, but it was the discovery of gold in New South Wales and Victoria in 1851 that launched large-scale migration to Australia and propelled a surge in urbanisation, with Australia becoming one of the most urbanised countries in the world by the beginning of the twentieth century (Laverty 1983).

Mining drove the growth of numerous smaller towns towards the base of the urban hierarchy, but also contributed to the expansion of major secondary centres in the form of industrial centres and provincial capital cities. Many of the smaller mining towns remain highly specialised and vulnerable to external demand shocks, although some of the larger centres, such as the gold town of Kalgoorlie in Western Australia, have achieved a degree of stability (Lawrie, Tonts and Plummer 2011).

Most recently, the adoption of so-called 'fly in, fly out' systems whereby labour is flown in on a weekly basis to remote regions such as the Pilbara in Western Australia (Lawrie, Tonts and Plummer 2011; Storey 2001) fosters a process of indirect urbanisation since some of the urban development effects take place in workers' towns of residence in addition to the mining towns themselves. In particular, this has reinforced the growth of Perth as the primary regional service centre and, to a lesser extent, Melbourne, which has been the centre of corporate control in Australian mining since the nineteenth century Victorian gold rush (Tonts and Taylor 2010, 2656).

Turning briefly to Latin America, silver mining was foundational to the economic and spatial organisation of the Spanish empire. Mines catalysed growth of the largest urban centres in the Americas (Dore 2000). The city of Potosi in present-day Bolivia was founded on silver mining, attaining a population of over 200,000 people at the turn of the seventeenth century. Moreover, cities such as Lima, Panama, Vera Cruz and Havana owed their growth to their positions on the official trade route from the mines to Spain, while Buenos Aires became the focus of the contraband silver trade.

Mining development in Latin America is also associated with the growth of secondary and regional cities in remoter areas. In the latter half of the twentieth century, Latin American governments sponsored major resource-based growth pole projects to develop remote regions, spurring the transformation of the Amazon region. Alongside large-scale state-led mining projects, Amazonia experienced an influx of artisanal prospectors or *garimpeiros* as part of a massive gold rush after a rise in the gold price in 1979, fuelling the rapid urbanisation of the region (Godfrey 1992).

Urban developmental processes in Sub-Saharan Africa: tracing mining's influence

While it is often remarked that Africa is the world's least urbanised continent, it should not be forgotten that archaeological evidence and explorers' accounts testify to the historical existence of densified populations on the African continent that rivalled the size and sophistication of European towns of the time (Anderson and Rathbone 2000; Freund 2007).[1] Coquery-Vidrovitch (2005) enumerates several necessary pre-conditions for historical delineation of urban settlement, namely: adequate staple food supply, trade and concentration of political authority, a specialised division of labour, class differentiation as well as money and monumental buildings.

We define urban settlements as populations of 10,000 people or above which are ethnically diverse, engaged in varied specialised economic activities rather than depending primarily on agriculture, and stratified by class in terms of different levels of income earning, asset holding and political power. In the early post-colonial decades, a number of typologies of African towns and cities were elaborated on the basis of the towns' economic functions, cultural features and political classifications (Vennetier 1976; O'Connor 1983). For purposes of comparing the nature of urban mining as opposed to urban non-mining settlements, our preference is for a simple four-fold classification that distinguishes functional origins and developmental dynamics, namely: first, centres of centralised power of state or religious authorities providing administrative governance; second, trade and transport centres, notably shipping ports and railway hubs; third, centres of industrial concentration; and, fourth, mining settlements, which some might consider as a subset of industrial centres, but their dependency on non-renewable mineral resources imparts a distinct form of temporality. Furthermore, the complex, bifurcated influences of artisanal as opposed to large-scale mining distinguish them further. In reality, urban settlements almost invariably are a blend of more than one of these four categories. However, for analytical purposes, it is helpful to disaggregate by function to perceive the varying character of mining settlements, thereby facilitating the periodisation of urban growth and settlement patterns historically into three main eras delineated below.

Empires, kingdoms and city-states: impetus of ancient long-distance gold trading

Mineral production and trade have had a significant influence on Sub-Saharan African urbanisation through history. Early urban settlements were generally associated with centralised kingdoms and/or important trading nodes for highly valued commodities including slaves, ivory and minerals, notably gold. Centres of religious and political authorities often merged. Early archaeological evidence of urbanisation exists along the Middle Nile at Meroë, the capital city of the Nubian kingdom of Kush from around 500 BCE to the fourth century CE (Welsby 1996). It is believed that the urban economy thrived on a rich array of exports including gold, ivory, ebony, semiprecious stones and exotic animal skins and feathers that were traded with the Hellenic city-states of the Mediterranean (Coquery-Vidrovitch 2005).

The rise of the Roman Empire, however, destabilised Kushitic urbanism. As the Mediterranean was reconstituted under Roman domination, trade between Europe and Africa shifted to the East African coastline.

The ancient period also witnessed the development of major urban mining and mineral-trading centres up-country. Monumental buildings and archaeological evidence of a large population concentration are found at Great Zimbabwe dating from the eleventh to mid-fifteenth century, closely associated with gold production. The gold was channelled into an Indian Ocean trade circuit with its main entrepôt in Kilwa (on the coast of present-day Tanzania) between the twelfth and fifteenth centuries and further south in the smaller ports of Ibo, Mozambique Island and Sofala in Mozambique (Bryceson et al., this volume). The towns declined in response to a diminishing supply of gold and the Portuguese ascendance as an Indian Ocean trading power in the sixteenth century.

Similarly, Timbuktu (in present-day Mali) has an illustrious history as a centre of long-distance trade and Islamic learning built upon commerce in gold and slaves. It coalesced as an urban locality in the twelfth century with the spread of Islam, attracting a cosmopolitan population composed of many ethnic groups. More generally, the Niger River region was famous for its large metropolitan cities including Gao, capital of the Songhai Empire, and Jenne, the Middle Niger delta's commercial centre (Saad 1983).

For centuries, alluvial gold produced in Ghana was traded along trans-Saharan routes destined for the towns of the Niger River (Gough and Yankson, this volume). The appearance of Portuguese traders in 1471 for gold and slave trading spurred gold reef mining. By the 1670s, as Timbuktu collapsed, and the Portuguese trade monopoly on the coast was superseded by competition between Dutch, Danish and English traders, considerable quantities of gold from Begho were directed south-wards to the Atlantic coast, becoming an important catalyst for the rapid rise of the Ashanti (Asante) kingdom in the latter part of the seventeenth century (Wilks 1993).

Colonial period: large-scale mining's imprint on urban Southern Africa

After the European scramble for Africa, economic and political influences took on new directions. The colonial governments structured their urban policies towards: first, limiting urban growth with the intention of keeping Africans in their rural home areas; second, facilitating labour control of African migration to the city (Rakodi 1986); and, third, segregating the races spatially within the city, using bye-laws and the *cordon sanitaire* under the pretext of ensuring European public health (Swanson 1977). Urban areas were, in effect, conceptualised as an alien environment that would contaminate individual Africans and corrode the communal fabric of tribal Africa. Their stated aim was to preserve the African agrarian way of life as illustrated by Felix Eboue, governor-general of French equatorial Africa, who argued in a circular dated 1941 for: 'the fixation of the African on the soil . . . in the midst of collective traditional institutions' (cited in Cooper 1996, 156)

The South African government and neighbouring colonial governments applied this thinking in the context of labour control for the expanding Southern African mining economy (Jeeves 1985; Mabin 1986; Moodie 1994; Harries 1994). During the latter half of the nineteenth century, colonial attitudes towards African urban settlement and residential accommodation for mining labour were closely entwined and codified in an urban racial segregation model that evolved in its most extreme form in South Africa. Given the importance of the segregation model for under-standing the impact of mining on African urbanisation during the nineteenth and twentieth centuries, it is worthwhile examining how it emerged.

The victory of the white settlers over the native Xhosa people in the 'last frontier' war of the late 1870s increased the number of Africans under colonial jurisdiction and crystallised a rationale for construction of a 'cheap labour reserve' (Kirk 1991). Many Europeans increasingly favoured withdrawing voting rights for propertied and employed Africans in the city to clear the way for the imposition of a comprehensive 'native policy' for Africans regardless of their class standing (ibid.). The Pass Laws and the Native Urban Areas Act of 1923 controlled the 'night residence' of Africans in urban areas. Apartheid legislation in 1952 added a battery of controls, with the labour bureau directing African workers to white employers.

Shedding light on how mining capital and the state shaped racially delineated space in South Africa's urban areas, Wolpe (1972) argues that male miners' low wages, short-term labour contracts and circular migration between their rural homes were posited on retaining the viability of the pre-capitalist mode of production. Miners' wives and children remained in the countryside relying primarily on subsistence agriculture, in effect subsidising corporate capitalist mining operations.

Mabin (1986) provides a valuable account of the closed mining compound model following the discovery of diamonds at Kimberley in 1869. Initially, diamond digging was pursued by large numbers of independent white small-scale claim holders. Blacks were not eligible to hold claims and were therefore restricted to working as hired labour, with their settlement dispersed and unplanned. The creation of the closed compound system only emerged at the Kimberley diamond mines in the 1870s. There was a meeting of minds that Africans should be segregated to facilitate control and to preclude 'black political dominance in the future' (Swanson 1977, 313).

At the early artisanal mining stage, the population at Kimberley reached 50,000. However, labour unrest and a decline in the diamond price resulted in diamond merchants seizing the opportunity to buy up claims. Concentration proceeded rapidly thereafter. By 1888, De Beers was only one of two large companies left, achieving a monopoly in 1889 (Mabin 1986).

As diamond production concentrated and mine shafts deepened, the pattern of diggers living close to their claims became obsolete. Open pit mines required increasing numbers of hired labourers. With growing concern over unrest in a multi-racial labour force and diamond theft, company managers began to favour building barracks alongside their pits to facilitate labour segregation and supervision. In 1884, De Beers took the lead in procuring cheap convict labour that they housed in closed compounds, which proved profitable and was thereafter followed by a closed compound for free labour (ibid.).

Kimberley nonetheless had large numbers of laid off and unemployed Africans who remained in the town in numbers roughly on a par with those of the black miners living in compounds. De Beers found their presence useful as a labour reserve, and the town's African housing became increasingly over-crowded as time progressed. After the Native (Urban Areas) Act of 1923 was passed, blacks were gradually forced to move into what became segregated, state-funded, municipally controlled housing. Both the mining compound and the segregated black townships became models for black urban residential housing in Southern Africa (ibid.).[2]

Gold mining on the Witwatersrand started in 1886, expanding rapidly to embrace a black mining labour force of 200,000 by 1910 (Jeeves 1985, 3). Thirty years later the total was 300,000, and by 1985 roughly 500,000 (Moodie 1994, 1). As Harrison and Zack (this volume) document, the foundations of Johannesburg were built on the Rand's gold mining operations. Johannesburg encapsulated South Africa's economic opportunities, contractions and racial oppression more than anywhere else, particularly in the apartheid period.

The implementation of the South African state's apartheid policy from 1948 onwards came at a time when the foundations of the national economy were shifting from gold and diamond mining to manufacturing and secondary industry (Wolpe 1972). Apartheid went beyond segregationist labour control and efforts to deflect the black population from urban settlement by seeking to establish a new system of nominally self-governing or 'independent' Bantustans in pre-existing and consolidated 'reserves'. Forced removal of surplus labour from the towns and white farms into the Bantustan reserves added to rural impoverishment. Meanwhile, segregated African townships were enforced by heavy investment in state security and police controls, setting in train a vicious cycle of black resistance and white repression.

In his article in this volume, Hugh Macmillan contextualises the Southern African historical debate about the relative costs of providing family wages and housing as opposed to low bachelor wages in the face of the inevitable dwindling supply of minerals. While South Africa witnessed the implementation of apartheid during the 1950s and 1960s, Northern Rhodesian (Zambian) copper mining firms invested in raising the productivity of their labour force with the provision of durable family housing and urban infrastructure as documented by Patience Mususa (this volume). Migrant workers to the Zambian Copperbelt were initially housed in camps, which had evolved into mining towns by the 1930s, accompanied by colonial administrative centres. In the 1950s and 1960s, living standards improved as a result of the colonial government's encouragement of trade unionism and the paternalistic approach of the large mining companies.

Post-colonial period: mineralising processes embedded in new urban social orders and disorders

As the wave of African national independence unfolded in the 1960s and 1970s, the racist constructs of the colonial division of labour were quickly dismantled. Capital cities of the newly emerged African nations attracted the main streams of urban migrants. The administrative and service sector positions that had been occupied by whites were Africanised. Family wages and salaries brought women migrants to the city in large numbers, evening out the sex ratio (Bryceson 1980). Most African countries' economies were based on agricultural exports, mainly produced by smallholder peasants, at a time in the 1960s when world agricultural commodity prices were high, favouring newly independent economies. One notable exception is Botswana, a Southern African country sparsely populated by agro-pastoralists, which experienced the windfall gain of diamond discovery at independence. As Thando Gwebu (this volume) describes, the government was sensitive to spending its financial resources judiciously from the outset.

In South Africa, the racial foundations of apartheid were attracting global protest and were challenged by domestic dissent regarding the black population's 'right to the city', directed at the inequities of black township relocations such as Soweto in Johannesburg and District Six in Cape Town. Increasing perforations in the circular migration system between home and mine appeared (Moodie 1994). During the mid-1970s mineworkers' wages started rising. Miners' union membership was accepted by the major gold-mining companies and some semblance of family reconstitution between female rural-based wives and male miners occurred through the migration of women to peri-urban areas of mining settlements following the repeal of the pass laws in 1986. At much the same time, by the 1990s, retired miners were sometimes opting not to return to their rural home areas and families, lacking sufficient capital to enjoy their status as male elders (Bank 1999).

In many parts of Sub-Saharan Africa, mining has become a source of conflict, reflecting the often irreconcilable claims of different groups for control over mineral wealth. Conflicts have multiplied in recent years as mining operations have expanded, often pitting mining corporations or the state against communities. At regional and national levels, liberation movements or rival groups fighting for territorial annexation and political power may seek control over mineral exploitation to further their cause (Le Billon 2001). According to Buijtenhuis (2000), peasants' wars directed at achieving African nationhood were eclipsed by predatory wars, where rival militarised groups sought state power, notably in Sierra Leone, Ivory Coast, Liberia, Democratic Republic of Congo and Angola. Mkandawire (2002) observes that rebel movement leaders increasingly have urban origins, with a tendency to choose roving rather than stationary military tactics due to lack of support from local rural peasantries. They targeted rich diamond areas, displacing the peasant farmers in their way. It is telling that minerals were not at issue in Mozambique's national liberation movement in the 1960s and 1970s, compared with Angola's and Sierra Leone's experience of civil war in the 1990s when 'blood diamonds' were used by rebel movements to fund conflict against established national governments (Rodrigues and Tavares, this volume; Maconachie, this volume). Finally, conflict can also arise between large-scale mining operations and artisanal miners, with the expansion of the former sometimes resulting in the displacement of the latter (Emel, Huber and Makane 2011; Gough and Yankson, this volume).

De Boeck's (1998, 2001) in-depth analysis of Congolese diamond smugglers' and diggers' forays into Angola reveals how the illicit trade in artisanal conflict minerals spurred rapid growth in a number of small towns, across the border in the DRC, considerably distanced from the country's older diamond centre of Mbuji Mayi, discussed by Patience Kabamba (this volume).[3] As Cristina Rodrigues and Ana Paula Tavares (this volume) document, much of Angola's urban growth pattern during the civil war depended on shifts in the fortunes of the government's MPLA (People's Movement for the Liberation of Angola) military forces versus UNITA (National Union for Total Independence of Angola). For the most part, the Angolan rural population of the Lundas attempted to seek safety in MPLA-held territories, while Congolese youth and Angolans from other parts of the country took the risk of migrating to UNITA-held areas of the mineral-rich Kwango River. Similarly in Sierra Leone, Roy Maconachie (this volume) recounts how rebel forces depended on artisanal diamond production, based upon youth labour recruitment (Le Billon and Levin 2009; Richards 1996). Most of the diamond-producing sites in very remote forest areas along rivers served as clandestine strongholds. However, the atrocities that the youth were ordered by their commanders to commit against the local society, sometimes against their kin and neighbours, caused estrangement if not a pariah status (Richards 1999). When hostilities officially ended in 2002, droves of the approximately 200,000 artisanal miners (Le Billon and Levin 2009) migrated to the capital city, swelling its population and posing a potential security threat. As Maconachie notes, their subsequent return to their rural home areas marks a dramatic development. In a spirit of physical and spiritual recovery, most have been welcomed rather than shunned, and availed agricultural land to make a new start (Richards 1996; Peters and Richards 1998).[4]

Africa's new mining boom: 2000 and beyond

Metals such as aluminium, copper, nickel, zinc and gold have experienced rapid price increases in recent years (Lawrie, Tonts and Plummer 2011), with gold prices soaring by over 600% since 2001 to a record high of US$1911 in August 2011, driven by investors' quest for a safe haven amidst financial crisis and recession (Blas 2011). Such price increases reflect the underlying volatility of commodity markets, following a prolonged period of price depression in the 1980s and 1990s (UNCTAD 2007), fuelled by speculation in financial markets and a new geography of demand with the growth of China playing a particularly important role.[5]

These price dynamics promise an unprecedented bonanza for four main sets of actors: the corporations who produce and sell the commodities, artisanal miners and the governments and populations of the countries in which the mineral wealth is located. For resource-rich developing countries, the new mining boom promises new development opportunities, seeming to reverse the historic decline in the terms of trade between their commodity exports and the manufactured goods and service of developed countries (Freudenburg 1992). Yet, given the notoriously asymmetrical nature of the relationship between transnational mining corporations and local communities, the boom over the long run may only serve to accentuate the contradictory role of mining as a vehicle for economic and social development (Bebbington et al. 2008).

This commodity boom follows the widespread liberalisation of mining codes in developing countries during the 1980s and 1990s, as part of a broader programme of economic and financial reform promoted by the World Bank and other international financial institutions (IFIs) (Bridge 2004). Based on comparative advantage and a sectoral approach to mining, the codes were designed to generate greater export earnings and attract foreign investment, privileging this approach over the adoption of an integrated development strategy with distributive goals (Campbell 2008).

Early indications are that the liberalisation policies are already favouring the interests of foreign capital over domestic interests and the artisanal mining sector[6] (Emel, Huber and Makene 2011; Hilson and Potter 2005). The relationship between the mining corporations and governments seems to have become more asymmetric (Bridge 2008). Liberalisation measures, which pointedly sought to reduce the role of governing states, have often undermined development capacities, making future reforms aimed at re-orienting the mining sector towards developmental goals more difficult (Campbell 2008).

The new mining boom serves to refocus attention on the resource curse thesis (Bridge 2008), which holds that reliance upon the exploitation of mineral resources is associated with poor economic performance. Critics can point to impending signs of 'Dutch disease' whereby mineral wealth pushes exchange rates and wages upwards, crippling the performance of non-mineral sectors such as agriculture and manu-facturing and creating an enclave economy divorced from other economic sectors.

Various commentators cite the failure of mining in poverty reduction (Freudenberg and Wilson 2002; Ross 2001). This is acknowledged by the mining industry itself, which tends to attribute the blame to the weak governance capacities of states, maintaining that mining remains good for economic growth (Bebbington et al. 2008). The lack of governance capacity reflects the neoliberal reforms of the 1980s and 1990s that reduced the role of the state and adopted a narrow sectoral approach to mining as a source of export revenue and a magnet for investment (Bryceson 2006; Campbell 2008).

In response to the negative effects of the resource curse thesis, IFIs have in recent years advanced what Arellano-Yangas (2008) calls a 'new natural resource agenda' that emphasises: decentralisation of government; greater citizen participation in decisions on how to spend mining revenue; and partnership between state agencies and mining companies. Time will tell how seriously IFIs and, most importantly, African national governments pursue these aims over the course of the mineralisa-tion of their economies.

Case studies of urban mining settlements in transformation

Framed by the preceding discussion of international patterns and processes and the schematic history of African urbanisation presented above, we now turn to mining and urbanisation trajectories in Sub-Saharan Africa considered in the case studies that follow. Our case study countries represent a range of national experiences arising from different temporal points of entry into mining, different configurations of artisanal and corporate large-scale mining, and different minerals. Gold and diamonds repeatedly surface as the key minerals, but copper (Zambia and Zimbabwe), chromite (Zimbabwe), uranium (Botswana) and other industrial minerals are included. Table 1 schematically indicates the range of case study diversity. While 'expanding' and 'contracting' mining sectors are self-explanatory, 'transitional' refers to circumstances in which mining is adjusting to new circumstances, be it a post-war situation, renewed investment or a new stage of the mining cycle.

Table 1. Thematic listing of case study mineral-producing countries.

	Form of mining		
Stage of mining	Large-scale	Large and artisanal	Artisanal
Expansion		Tanzania, Democratic Republic of Congo (DRC)	Zimbabwe
Transitional	Zambia, Botswana, Angola	Ghana	
Contracting	South Africa, Zimbabwe	Sierra Leone	Angola

Mining and urban systems

How mining shapes urbanisation at the level of national urban systems as well as within individual urban settlements in Africa is a relatively unknown arena. Obviously, mining attracts migrants with the expectation of more remunerative livelihood opportunities, but the question of how people move to, settle in and sustain a livelihood in the face of the roller coaster of global mineral markets remains largely unanswered (though see Rajak 2012; Walsh 2012). From recent observations of the current mining boom, settlement patterns of artisanal and large-scale mining differ temporally, spatially and infrastructurally. Artisanal mining tends to appear first and is most often associated with the rapid growth of small towns and regional centres, which commonly wanes as the mineral supply is depleted or unreachable with artisanal technology (Jønsson and Bryceson 2009; Bryceson et al., this volume). Large-scale mining, on the other hand, is likely to support larger *in situ* urban settlements as well as contributing to the growth of capital cities and service centres (Tonts and Taylor 2010).

At the time of South Africa's dramatic transformation into a country dominated by mining, the port cities of Cape Town and Port Elizabeth were the most populous urban settlements. By 1900, Johannesburg's growth had overtaken both of those settlements, and it has retained its supremacy; although Durban and Cape Town are a close second and third.[7] Considering South Africa's largest cities, one would therefore assume that urban primacy is not pronounced in South Africa (Figure 1). However, Johannesburg and Pretoria, the South African political capital, have been administratively conjoined since 1994 in the new Gauteng province, which also includes Ekurhuleni, comprised of a string of former Rand mining towns.[8] The name 'Gauteng' derives from the Sesotho word for gold, *kgauta* (possibly derived from the Afrikaans word *goud*), and the locative suffix -eng, and means 'place of gold'. In effect, Gauteng forms an extended urban megalopolis with a population of 8.8 million people during the 2001 census (almost 20% of the South African population). In this context, South Africa represents a case of extreme urban primacy in which the political and economic capitals have merged with local municipalities to form one enormous urban conurbation derived from gold mining.

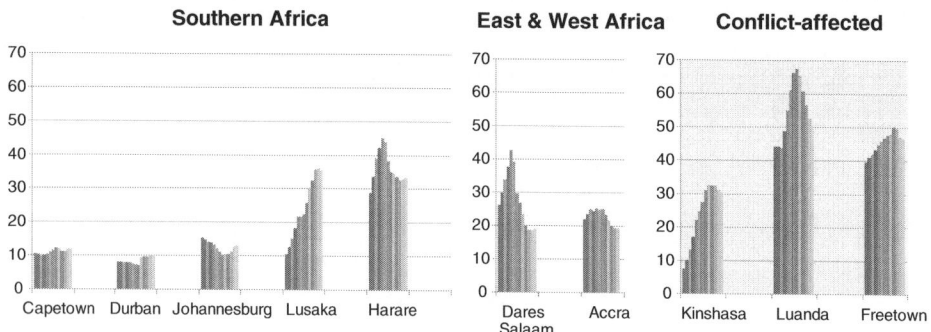

Figure 1. Urban primacy patterns (percentage of total urban populations), 1950–2010. Individual columns of each city represent 5 year periods ranging between 1950 and 2010. Source: United Nations Population Division 2004.

Zambian towns are quite different. Out of the country's eight most populous urban settlements after Lusaka, six are or were originally mining towns; five are located in the Copperbelt[9] as well as zinc and lead-producing Kabwe. As Mususa (this volume) documents, declining copper prices have spelled misery for Copperbelt residents, resulting in the contraction of Copperbelt towns, while Lusaka grew (Figure 1). Thus, contrary to the tendency for capital cities' urban primacy to decline in the decades following national independence, Lusaka gained primacy, as observed by Macmillan (this volume).

In Botswana, Gwebu (this volume) traces how urbanisation has been dominated by large-scale mining, providing the economic basis of most major towns outside of the capital city, Gaborone. Similarly, Amin Kamete (this volume) documents how mining has exerted a strong influence on urbanisation in Zimbabwe, underpinning the growth of numerous urban centres, although the country's two largest cities, Harare and Bulawayo, are not directly associated with rich mineral deposits in their immediate vicinity. Nonetheless, the concentration of corporate headquarters has contributed to the growth of Harare, although urban primacy has been on the decline since independence in the 1970s (Figure 1).

Bryceson et al. (this volume) show that small towns in Tanzania's mining districts have grown more rapidly than those in non-mining regions during the 1980s and 1990s, especially south of Lake Victoria in the so-called 'ring of gold'. This is likely to have contributed to Dar es Salaam's declining urban primacy (Figure 1). In Ghana, too, Gough and Yankson (this volume) indicate that mining has contributed largely to the growth of secondary centres and smaller towns.

Kinshasa's urban primacy in the DRC is very striking, dwarfing Lubumbashi, the country's second largest city. It is noteworthy that the next eight towns in the urban hierarchy are in areas of mineral wealth, including Mbuji Mayi.[10] Similarly, in 2004 Freetown's population was over five times that of the second largest city. Its primacy continually increased during the civil war and has only now started to abate in the post-war period. Finally, Angola, like Mozambique, had an urban colonial heritage, which accorded primacy to the capital city Luanda and thereafter was fuelled by the disruptions of civil war that caused people to flee to Luanda, as explained by Rodrigues and Tavares (this volume).

Mining settlement and growth dynamics

At a more localised level, mining has given birth to urban centres as well as propelled their growth over the past two decades of neo-liberal policy implementation. The current global mining boom is fuelling new patterns of settlement in and around key 'rush' sites, as exemplified by rapid in-migration to mine strike sites in Tanzania (Bryceson et al., this volume). The pattern of initiation, growth and decline typically experienced by mining towns has a cyclical character driven by the interaction between resource availability and exploitation, capital accumulation, infrastructure development and labour and population flows.

Bryceson and Jønsson (2010) provide a stylised account of the process of settlement for Tanzania. In the pre-phase, minerals have yet to be discovered or have been discovered but only mined by a few local people who still rely primarily on agricultural production. This state can go on indefinitely but rising global mineral prices trigger the first mining phase, which is characterised by a rapid influx of artisanal miners; closely followed by a second phase, involving a wave of non-mining migrants to work in trade and services. Their combined, multi-skilled in-migration effects urban transformation. The sex ratios of the settlement equalise as local economic multipliers develop alongside increasing mineral production. Eventually, this gives way to a third phase whereby miners' labour reaches a stage of diminishing returns, when mineral extraction becomes too difficult and costly in the light of available technology and international demand conditions. These circumstances reduce the rate of in-migration, leading to Phase 4 when the rate of out-migration exceeds the rate of in-migration.

A related pattern is that of artisanal mining displacement that is most likely to occur as a settlement passes from the second to the third phase. The mineral supply is still very rich but digging has descended to a depth that artisanal miners cannot reach with their technology. This often leads to the expansion of large-scale corporate mining, more often than not under foreign ownership, management and control. Global interests supersede local interests. This is a critical point for contestation (Hilson and Yakovleva 2007). Many African governments are too inexperienced, weak or corrupt to broker deals to ensure sufficient benefits for the local mining population and the region and country at large.

As mining towns grow, their economic character evolves. Harrison and Zack's article (this volume) centres on Johannesburg, tracing its transformation from a mining camp into the primary mining-based metropolis in Southern Africa. The scale of the gold industry around Johannesburg meant that it exerted major 'backwash' effects on neighbouring countries in Southern Africa and beyond by drawing in foreign labour, in addition to the South African workers who were subjected to coercive methods or recruitment.

Gough and Yankson (this volume) describe how modern gold mining in Ghana expanded greatly in the so-called 'Jungle Boom' of 1901–1902, with urbanisation occurring along the south coast where maritime trade was centred and the main mineral resources located. Mining not only continued to support the development of established towns such as Obuasi and Tarkwa, but has also fuelled the growth of new settlements in the Northern region based on open-cast extraction of gold by artisanal miners (Hilson and Ackah-Baidoo 2011).

Maconachie (this volume) traces the growth of diamond mining in Kono district in the Eastern Province of Sierra Leone starting with large-scale in-migration in the 1940s and 1950s. Crucially, this process encouraged the growth of agriculture, retail and other services to supply the mining population, particularly in burgeoning towns such as Kenema and Koidu.[11] Since the cessation of the civil war, artisanal mining has declined as alluvial diamond deposits have become increasingly worked out, in many cases spurring a return to agriculture.

In the DRC, on the other hand, agriculture and mining have different regional trajectories. In Butembo's rich agricultural environs, commercial production of coffee and subsistence food production have continued whereas agrarian options in Mbuji-Mayi are unlikely to be looked upon with much favour (Kabamba, this volume). Parks (2011) expresses doubt about smallholder agriculture's potential as an exit option in many mining areas of the DRC given artisanal miners' loss of agricultural skills, land access and the relatively poor remuneration from farming compared with what they have grown accustomed to in mining. Those facing dwindling income from depleting pits are more likely to seek trade and service opportunities.

In Zambia, Mususa (this volume) identifies a transition from mining to agriculture, which is primarily *in situ*. The unexpected opportunity that the government-owned mines offered to laid-off workers to buy the houses they had occupied while working helped to stave off migration to Lusaka and elsewhere. Those who made the house purchases thereafter tended to try to make ends meet by turning to agriculture, hence her observation that the towns have become like villages.

In Zimbabwe, the same issue arose, but the availability of mine housing in the face of a mineral slump coincided with traumatic displacement through an urban 'clean-up' campaign in Harare and other towns, which affected approximately 700,000 people. The impact was compounded by 300,000 African farmworkers' evictions from white-owned farms. Kamete (this volume) shows how mining company 'ghost towns' became the refuge of the displaced from these campaigns, particularly foreign workers who had no rural home areas to return to in Zimbabwe. Their housing problem was addressed but viable livelihoods remained a struggle. Many turned to very low-paid sub-contracted mining. Poverty and hunger reached levels necessitating the distribution of food aid.

Sense of urban mining place

Here, we dwell on two main characteristics of urban mining places: their cosmopolitanism and their temporality, which tend to mark them as distinct from other urban sites within the nation-state. Cosmopolitan mixing of people from a wide array of ethnic groups is one of the defining features of urban life. But it is not the mere physical co-existence of multi-ethnic people that matters. As Appiah (2007) argues, the 'diversity principle', a pervasive acceptance of ethnic diversity open to relational ties and moral responsibility regardless of ethnic origins is vital to urban social coherence. Key here is the interplay between individual urban identity formation and the congealing urban social collectivity over time.

Artisanal mining, operating at lower levels of remuneration, provides more labour absorption generally, giving rise to far larger numbers of urban migrants than large-scale mining with its greater capital. Artisanal mining settlements tend to be embryonic cosmopolitan locations from the moment of their origin, especially at mineral rush sites where miners from near and very far away amass. Artisanal mining sites in Tanzania and Ghana have been noted for their ethnic diversity with a considerable sprinkling of foreign nationals as well (Jønsson and Bryceson 2009; Gough and Yankson, this volume). Artisanal mining by its very nature of involving groups of people in collective work of an uncertain and dangerous nature demands trust. Bryceson and Jønsson (2010) found that outside of miners' first apprenticeship mining site, Tanzanian artisanal miners judged each other on the basis of mining skills and competence instead of place of origin and tribal affiliation.[12]

By contrast, the Southern African mining compound model operated largely on the basis of tribal groups. Mine companies recruited widely within Southern Africa, but recruitment practices, work teams and residential arrangements divided men along ethnic lines. Senior men acted as 'boss boy' team leaders in the mines and the *indunas*, management-appointed black supervisors, were charged with keeping order in the compound (Moodie 1994). The localisms of speaking a South African vernacular language, traditional tribal values and a sense of rural home were more likely to persist in such settings.[13] Already in the 1910s, South African mines sought to recruit so-called 'tropical' labour from Malawi and Mozambique (Jeeves 1985). Mine management and the South African government's divide and rule tactics made for fractious work relations amongst different ethnic groups and between South Africans and foreign nationals rather than harmonious cosmopolitanism (Moodie 1994). In the new South Africa, these labour practices have been abandoned but one need only look at post-apartheid South African censuses to see the delineation of ethnic, race and language groups still meticulously recorded as essential components of municipalities and the nation state.

In the historical Southern African model of urban deterrence, the mining camp, compound and township were residential constructions moulded to racially restrict Africans from permanent urban settlement. As Macmillan (this volume) points out, there has been a long and ongoing debate about whether Africans were surmounting these restrictions and permanently settling in urban areas, claiming their right to the city as urban residents, or maintaining a pattern of temporary, circular migration and persistent links with the countryside (Gluckman 1961; Mabin 1989; Ferguson 1999; Potts 2010).

South African census statistics indicate that racial segregation and apartheid may have slowed, but not stopped, the process of African urbanisation. Growth rates for African populations in urban areas exceeded that of whites from the 1920s onwards (Beall, Crankshaw and Parnell 2006). Tellingly, Southern African mining countries had higher levels of urbanisation than elsewhere on the continent by the 1960s. Urban growth, permanent African urban settlement and what colonial regimes referred to as 'detribalisation' progressed, despite the system of rigid controls designed to preclude them. Gradually, the sex ratio equalised as women found ways of moving close to the mines where their husbands worked (Moodie 1994).

In Northern Rhodesia, bachelor wages and circular migration were replaced with family housing and wage packets at the mines, though not until the late colonial era. Family formation and reunion were at last sanctioned. Mususa (this volume) describes the Copperbelt's investment in housing and infrastructural building. Alongside the bricks and mortar of African family dwellings, residents were constructing urban mining identities, cosmopolitan cultural styles and new community values. As those permanently settled had families and raised their children in the urban context, rural idioms sometimes became quaint. Nonetheless, some second and third generation residents of the Copperbelt were forced to resort to the agricultural livelihood of their rural forefathers in the dark days of recession during the late 1990s, as some parts of the Copperbelt became 'like a village'.

These experiences reveal how a sense of urban mining place necessarily entails the flexibility to adapt to minerals' global price fluctuations as well as the inevitability of declining mineral supplies as non-renewable resources. This is true for large-scale mining towns as well as artisanal mining settlements. In Botswana, as the copper-nickel deposits in Selebi Phikwe became uneconomic to exploit, the government endeavoured to bring about an industrial and commercial transformation of the town.

Similarly, artisanal miners working without the security of fixed term contracts generally have an acute appreciation of the temporality of urban mining places. In Ghana and Tanzania, miners who have managed to save money over time are likely to invest in building houses, generally in a nearby trading settlement or the regional capital, rather than the mining site, precisely because of their sense of urban mining place (Gough and Yankson, this volume; Bryceson and Jønsson 2010).

Artisanal mining settlements do not necessarily become ghost towns when the minerals are depleted. They are likely to survive as urban service centres if they are located on a main road, not constrained by water shortages and offer a service infrastructure that is superior to rural settlements in the surrounding area, particularly those close to new mineral rush sites. Having been substantial settlements in their mining heyday, they usually have considerable educational, health and other amenities. And as Kamete (this volume) documents, towns such as the abandoned corporate mining town of Mhangura, with its crumbling housing and service structure, nonetheless became a haven for Zimbabwe's politically displaced foreign workers, some of whom took up tribute mining as a last resort.

Mining's impact on urban development, welfare and poverty

Mining creates stark economic variations in wealth and poverty, which has taken racial and sometimes ethnic patterns, as well as creating wealth disparities and tensions between rural agricultural and mining communities in various instances (for example, Bryceson et al. 2011). What is striking is that welfare in formal urban mining settlements has generally been an outcome of state and corporate mining policies. Trade unionism emerged among African mineworkers in South Africa, Ghana and Zambia in the 1940s, but the South African union was crushed after a national strike in 1946. The present National Union of Mineworkers was not established until the 1980s. Without trade unions to fight for their welfare, workers in large-scale mines had been left to the mercy of the government's and the mining corporations' attitudes towards workers' welfare.[14]

The development of corporate social responsibility is based on mining companies' efforts to secure a 'social licence to operate' which entails utilising environmental and social standards to predict and limit negative external impacts and curb in-migration (Shankleman 2011). While expunging the South African compound model and embracing what is now known as the Ghanaian open model of mineworkers residing in the adjoining towns, some of the big international companies have provided major infrastructural improvements, including water installations, phone masts, health dispensaries and improved roads for the benefit of town residents generally.[15] Chinese mining investment often has gone further with the provision of major installations like pipelines, transport and shipping installations and sports stadiums, which tend to benefit Africa's capital cities and ports.

Gwebu (this volume) assesses the significance of diamond revenues for the welfare of Botswana as a whole. The population rapidly transformed from primarily rural cattle keepers to urban dwellers in multi-occupational settings within one generation. Special attention has been paid to the equitable regional division of the benefits. Welfare indicators rose accordingly with regard to access to the basic needs of food, water and housing, reflected in its middle-income status according to the World Bank's classification. Whilst it represents the exception to the rule in Sub-Saharan Africa, the Botswanan case underlines the need for a strong development state with the institutional capacities to harness mining revenues for development.

The experience of the DRC, on the other hand, provides a starkly contrasting illustration of the resource curse thesis, with its massive mineral wealth matched only by its status as one of the least developed countries on earth. Against a history of plunder and conflict, Patience Kabamba (this volume) draws a distinction between processes of urbanisation in the relatively prosperous gold city of Butembo and the impoverished diamond city of Mbuji-Mayi. He argues that the cultural norms of Nande gold traders investing in Butembo housing and infrastructure encourage local embeddedness, whereas the Lunda diamond traders are socially fragmented and reliant on external patronage, fostering widespread insecurity and poverty in Mbuji Mayi and the Lunda region generally.

As Kamete (this volume) shows, the role of nation-states in mediating between global and local forces is not always positive, as illustrated by the Zimbabwean state's ruinous economic policies and political repression. People displaced from rural and urban areas by the government's land reform and 'urban clean-up' campaigns in the 2000s settled in former mining towns, resulting in both demographic revival and desperate impoverishment. The fate of the Zimbabwean mining towns provides a unique mutation of the mining settlement cycle whereby decline is followed by a new 'growth' phase based on forced migration without any basis in capital accumulation.

Another important set of dynamics has been at work on the Zambian Copperbelt, where a system of welfare capitalism that had been established in the 1950s finally collapsed with preparations for privatisation in the late 1990s, resulting in high levels of poverty and the informalisation of livelihoods alongside subsistence agricultural production and some artisanal mining. Mususa (this volume) argues that the decline of mining has led to higher levels of poverty within some mining settlements, though there has been a partial recovery on the old Copperbelt in the last decade, as well as new mining developments in the North Western Province. Foreign investors, from Canada, India and elsewhere, have injected large amounts of capital, but have not been prepared to revive the welfare systems, including schools and clinics.

The past is even more difficult to reconcile in terms of the loss of life and misery that conflict mineral production has inflicted on the African continent. Conflict mineral exploitation is categorically devoid of welfare benefits for those subject to its machinations. As Maconachie (this volume) documents, the rebel-affected diamond-producing areas of Sierra Leone contain some of the worst poverty in the country. Kono district is still undergoing a painful process of reconciliation with hopeful signs of reduced tension.

Finally, there is the issue of welfare for miners who seize the opportunity to improve their standard of living and invest for the future. Significantly, evidence from Tanzania, Angola and Ghana suggests that artisanal miners have a tendency to avoid building permanent houses on mineral rush sites, preferring to invest in modern housing in nearby trading centres or regional towns (Bryceson and Jønsson 2010). This tends to reflect artisanal miners' acute appreciation of the temporality of urban mining places. In this way, mining spurs an indirect and locationally deflected urbanisation.

Conclusion

Our introductory overview has been drawn with broad-brush strokes. Much of the interplay of mining and urbanisation is unrecorded, buried in the mists of pre-colonial history, then considered self-evident and not worth mentioning in colonial records and similarly overlooked in bureaucratic reports of post-colonial governments in which mining and urban ministries occupied different mindsets. Detailed case studies of the intersection of the two are now emerging, as demonstrated by the national case studies in this volume and various recent publications written primarily by anthropologists based on first-hand accounts of artisanal and corporate mining.[16] As the number of Sub-Saharan mining investments by small and large-scale mining interests proliferate over the coming decades, there is a need for more local studies and macro analyses of mineral production and trade figures in relation to urban population data.

While this review has identified a number of thematic threads that hint at recurring interactive patterns, there is no grand narrative or iron law whereby mining dictates specific urban outcomes. Instead, the connection appears to be a matter of historical contingency in the context of global mineral demand. Over the broad sweep of African trade history, gold has played a significant role, with its enduring intrinsic value and demand through the ages as a luxury commodity. Beginning in the late nineteenth century and expanding throughout the twentieth century, diamonds gained in importance, alongside industrial minerals like copper, tin, zinc, uranium, platinum and coal and more recently highly valued rare earth minerals including columbite-tantalite (coltan).

Mineral exploitation through time has been moulded by prevailing technology and institutional arrangements. Insofar as Africa's ancient history is discernible, gold was a major luxury whose production and trade was associated with African kingdoms and chiefs. Already in the thirteenth century, there were African gold mining sites connected to Arabian, Asian and European markets, via East African Swahili coastal city-states. Elite, enclave cities for trade and large populous, but probably makeshift, production centres formed the urban mining landscape of that era.

From chiefs and city-states, the organisation of mineral production and trade gradually shifted over the centuries to the coordinated exercise of corporate and state power. In the late nineteenth and early twentieth century, as European colonial rule engulfed the continent, a racial policy of perverse urbanisation based upon African rural–urban circular migration, mining compound residences and Bantustans was instituted in South Africa. Segregation and apartheid did not curb urbanisation as intended, but inflicted dislocation and damage on African family coherence and fostered pervasive racial discrimination in urban society.

The late twentieth century witnessed widespread deagrarianisation of African economies and African governments' adoption of neo-liberal policies, which propelled rural Africans to search for remunerative forms of income (Bryceson 2000). Large numbers gravitated towards artisanal mining with the expectation of high earnings. This opportunity was also snatched by rebel armies seeking sources of finance, notably from diamonds, for waging civil war, which placed mining in the thick of recent African history. Over the past decade, surging world commodity prices and liberalised investment regimes have led to a resurgence of international capital's investment in mining in Africa; however, artisanal mining tends to involve far more people in urbanising processes than large-scale mining. Artisanal miners are also more mobile in moving between sites in response to mineral depletion.

At present, housing in mining settlements encompasses what is either provided to or afforded by large-scale mine workers, on one hand, and artisanal miners' generally informal and often makeshift housing, on the other. Welfare differences between the two cannot be encapsulated in a simple dichotomy between adequate and inadequate. Miners employed in large-scale mines may face employers' continued reluctance to provide for long-term family housing, while artisanal miners' seemingly poor makeshift circumstances may be a pragmatic choice while they save their earnings to build a house elsewhere.

African mining and urbanisation narratives, from the beginning of the eureka moment in mineral discovery, are about temporality and the need for 'digging deeper' or 'moving on'. Inevitably, this has a direct effect on the nature and spatial pattern of urban settlement. Eventually, either the mine production or mineral trading withers and dies, with its residents scattering elsewhere, or alternatively other *in situ* livelihood activities arising, or some combination of the two occurring. What makes the story of African mining and urbanisation so diverse in settlement trajectories and erratic in livelihood and welfare paths is how the temporal and spatial limitations of mineral availability intersect with miners' work and residential rights. Family welfare largely depends on miners receiving sufficient earnings and scope to establish their families' residence locationally alongside their mine work. This was not the case under apartheid and has various degrees of resolution at present in artisanal and large-scale mining, given prevailing political power structures and the inevitability of mineral depletion.

Finally, we return to our earlier concept of mineralising urbanisation. We have argued that many countries on the African continent are currently experiencing successive localised mining booms and busts catalysing profound local, regional and national effects on urban settlement. While artisanal and corporate mining impact on urban change in markedly different ways, the effects are nonetheless interwoven as revealed in the articles in this collection. There is a profusion of temporal uncertainties in relation to mineral discovery, access and depletion, fluctuating global mineral prices, changing corporate management ownership and policy shifts, and government intervention regarding mineral access. After the certainty and euphoria of mineral discovery comes the uncertainty of settlement trajectories with highly indeterminate outcomes. As African large-scale and artisanal mineral output accelerates, the ensuing complex changes in settlement patterns are integral to profound transformation of African national political economies and societies. The interaction of mining and urbanisation needs to be documented, understood and acted upon to ensure a more productive, just and tranquil future.

Acknowledgements

This special issue derives from papers presented at an international workshop held in Bagamoyo, Tanzania in February 2011. We are grateful to the United Kingdom's Department of International Development (DfID) and the Economic and Social Research Council (ESRC RES-167-25-0488) for their financial support of the University of Glasgow's Urban Growth and Poverty in Mining Africa (UPIMA) research programme which sponsored the workshop.

Notes

1. In pre-colonial history, urban areas were likely to consist of populations of between 5000 and 20,000, sometimes more. There are virtually no *in situ* records of population totals. A handful of foreign visitors wrote travel journals where they were apt to estimate the population of large, impressive settlements, but no one operated with notions of critical population levels.
2. Freund (1981, 79) notes that the Nigerian Chamber of Mines requested a government organised labour bureau modelled after the Rhodesian Native Labour Bureau in 1911 with the intention of reducing the cost of wages. It met with rejection on the grounds that the government felt it would interfere with peasant export crop production.
3. It is estimated that Angola's rebel force UNITA was earning $300–350 million per year (*Financial Times*, 3 May 1996, cited in De Boeck 1998, 782).
4. Note that the majority of internally displaced persons were 'ordinary' rural citizens caught up in the conflict, an estimated 500,000 farming families. Their entry into Freetown and return to home areas represented a massive population shift exceeding that of ex-combatants.
5. China accounted for 29%, 66% and 25%, respectively, of the growth in demand for oil, copper and nickel in 2005 (UNCTAD 2007, 89).
6. For example, Le Billon and Levin (2009, 701) report: 'Operation Brilhante (Shining) – facilitated the further industrialisation of the diamond sector and brought the ratio of artisanal to industrial production in value from a quarter to an eighth between 2002 and 2006. Angola has achieved the highest fiscal ratio among conflict-affected countries and one of the highest growth rates in tax receipts over the past five years. Yet cronyism is rampant in the mining industry and the KPCS [Kimberley Process Certification Scheme] itself is used as an excuse for criminalising artisanal miners'.
7. In the 2001 South African census the population totals were: Johannesburg 3.2 million; Durban 3.1 million; Capetown 2.9; and Port Elizabeth 1.0 million.
8. In addition to the metropolitan municipalities of Johannesburg, Tshwane and Ekuhuleni, there are two district municipalities divided into seven further local municipalities.
9. Kitwe, Ndola (now an administrative, commercial and mine service centre), Chingola, Mufulira, Luanshya.
10. The DRC has not had a census since the 1980s but 2004 estimates based on the United Nations MONUC-DPKO GIS Unit listing its principal cities in descending population are: Lubumbashi 1.28 million (copper); Mbuji-Mayi 1.21 million (diamonds); Kananga 0.72 million (copper, cobalt); Kisangani 0.68 million (diamonds); Bukavu 0.47 million (gold and tin); Kolwezi 0.46 million (copper); Likasi 0.37 million (copper and cobalt) and Tshikapa 0.37 million (diamonds).
11. As of 2004, Kenema (population 128,402) and Koidu (population 82,899) are the third and fourth largest urban concentrations in Sierra Leone.
12. Grätz (2009) describes similar circumstances in artisanal gold mining settlements of Northern Benin.
13. It should be noted however that the official language of the mines was Fanakalo, a mine pidgin (based on Zulu), compulsory underground for all black workers and white supervisors for safety reasons.
14. As we go to press (August 2012), the integrity of the National Union of Mineworkers (NUM) to fight for the welfare of South African miners was being challenged in the aftermath of a police massacre of striking miners at the South African Lonmin platinum mine near Marikana, killing 34 and injuring 78 miners. Julius Malema, an ex-African National Congress (ANC) political leader, blamed the South African President, Jacob Zuma, and the ruling ANC party. Malema announced his support for the formation of a militant union in place of the ANC-aligned NUM ('Expelled ANC youth leader calls for state takeover of mines in wake of massacre', London *Guardian*, 19 August, 2012; 'South Africa's Mines: In the pits', *The Economist*, 25 August, 2012).
15. Interview with Mining Officer Mr Joseph Mangilima at the Anglo-Gold Ashanti-owned Geita Gold Mine in Geita, Tanzania, 4 September 2011.
16. See Carstens 2001; De Boeck 2008; Duffy 2007; Luning 2006; Verhoeve 2004; Vlassenroot and Büscher 2009; Walsh 2003, 2012; Werthman 2009.

Notes on contributors

Deborah Fahy Bryceson is a Reader in the School of Geographical and Earth Sciences at the University of Glasgow. She has published on the theme of African deagrarianisation and livelihood diversification including *Farewell to Farms: De-agrarianisation and Employment in Africa* (Ashgate 1997); *Disappearing peasantries: Rural Labour in Africa, Asia and Latin America* (Intermediate Technology Publications 2000); *African Urban Economies* (Palgrave Macmillan 2006); and *How Africa Works: Occupational Change, Identity and Morality in Africa* (Practical Action Publishing 2010). She currently coordinates the Urbanisation and Poverty in Mining Africa (UPIMA) research programme. She can be contacted at: dfbryceson@bryceson.net

Danny MacKinnon is a Senior Research Fellow in the School of Geographical and Earth Sciences at the University of Glasgow. He is an economic geographer with research interests in urban and regional development, the adaptation of cities and regions to global processes and forms of regional governance and devolution. He has published extensively on these themes in a range of international journals such *Economic Geography, Progress in Human Geography, Journal of Economic Geography* and *Regional Studies*. He can be contacted at: daniel.mackinnon@glasgow.ac.uk

References

Anderson, D.M., and R. Rathbone. 2000. Urban Africa: Histories in the making. In *Africa's urban past*, ed. D.M. Anderson and R. Rathbone, 1–17. Oxford: James Currey.

Appiah, K.A. 2007. *The ethics of identity*. Princeton, NJ: Princeton University Press.

Arellano-Yanguas, J. 2008. A thoroughly modern resource curse? The new natural resource policy agenda and the mining revival in Peru. IDS Working Paper no 300. Brighton: Institute of Development Studies.

Bank, L. 1999. Men with cookers: Transformations of migrant culture, domesticity and identity in Duncan village, East London. *Journal of Southern African Studies* 25, no. 3: 393–416.

Barnes, T.J., R. Hayter, and E. Hay. 2001. Stormy weather: Cyclones, Harold Innis and Port Alberni, BC. *Environment and Planning A* 33: 2127–47.

Beall, J., O. Crankshaw, and S. Parnell. 2006. A matter of timing: Migration and housing access in metropolitan Johannesburg. In *African urban economies: Viability, vitality or vitiation?*, ed. D.F. Bryceson and D. Potts, 233–53. London: Palgrave Macmillan.

Bebbington, A., L. Hinojosa, D.H. Bebbington, M.L. Burneo, and X. Warnaars. 2008. Contention and ambiguity: Mining and possibilities of development. *Development and Change* 39, no. 6: 887–914.

Blas, J. 2011. Wild week as gold touches record before plunging. *Financial Times*, August 26.

Brands, H.W. 2002. *The age of gold: The California gold rush and the new American dream*. New York and London: Doubleday.

Bridge, G. 2004. Mapping the bonanza: Geographies of mining investment in an era of neoliberal reform. *The Professional Geographer* 56: 406–21.

Bridge, G. 2008. Global production networks and the extractive sector: Governing resource-based development. *Journal of Economic Geography* 8: 389–419.

Bryceson, D.F. 1980. The proletarianisation of women in Tanzania. *Review of African Political Economy* 17: 4–27.

Bryceson, D.F. 2000. Disappearing peasantries? Rural labour redundancy in the neo-liberal era and beyond. In *Disappearing peasantries?*, ed. D.F. Bryceson, C. Kay, and J. Mooij, 299–326. London: Intermediate Technology Publications.

Bryceson, D.F. 2006. African urban economies: Searching for the sources of sustenance. In *African urban economies: Viability, vitality or vitiation?*, ed. D.F. Bryceson and D. Potts, 39–66. London: Palgrave Macmillan.

Bryceson, D.F., and J.B. Jønsson. 2010. Gold digging careers in rural Africa: Small-scale miners' livelihood choices. *World Development* 38, no. 3: 379–82.

Bryceson, D.F., J.B. Jønsson, and R. Sherrington. 2010. Miners' magic: Artisanal mining, the albino fetish and murder in Tanzania. *Journal of Modern African Studies* 48, no. 3: 353–82.

Bryceson, D.F., and R. Mwaipopo. 2010. Rural–urban transitions in Tanzania's Northwest mining frontier, 2010. In *Rural–urban dynamics: Livelihoods, mobility and markets in African and Asian frontiers*, ed. J. Agergaard, N. Fold, and K.V. Gough, 158–74. London: Routledge.

Buijtenhuis, R. 2000. Peasant wars in Africa: Gone with the wind. In *Disappearing peasantries? Rural labour in Africa, Asia and Latin America*, ed. D.F. Bryceson, C. Kay, and J. Mooij, 112–22. London: Intermediate Technology Publications.

Campbell, B. 2008. Reform processes in Africa: Issues and trends. Paper presented at the 2nd International Study Group (ISC) Meeting, United Nations Commission for Africa, May 19–21, 2008, Addis Ababa.

Carstens, P. 2001. *In the company of diamonds: De Beers, Kleinzee and the control of a town.* Athens, OH: Ohio University Press.

Cooper, F. 1996. *Decolonisation and African society: The labor question in French and British Africa.* Cambridge: Cambridge University Press.

Coquery-Vidrovitch, C. 2005. *The history of African cities south of the Sahara.* Princeton, NJ: Markus Weiner Publishers.

De Boeck, F. 1998. Domesticating diamonds and dollars: Identity, expenditure and sharing in Southwestern Zaire (1984–1997). *Development and Change* 29: 777–810.

De Boeck, F. 2001. Garimpeiro worlds: Digging, dying and 'hunting' for diamonds in Angola. *Review of African Political Economy* 28, no. 90: 549–62.

De Boeck, F. 2008. Diamonds without borders: A short history of diamond digging and smuggling on the border between the Democratic Republic of Congo and Angola (1980–2008). In *Artisanal diamond mining: Perspectives and challenges*, ed. K. Vlassenroot and S. van Bockstael, 41–55. Gent: Academia Press.

Dore, E. 2000. Environment and society: Long-term trends in Latin American mining. *Environment and History* 6: 1–29.

Duffy, R. 2007. Gemstone mining in Madagascar: Transnational networks, criminalisation and global integration. *Journal of Modern African Studies* 46, no. 2: 185–206.

Emel, J., M. Huber, and M.H. Makene. 2011. Extracting sovereignty: Capital, territory and gold mining in Tanzania. *Political Geography* 20: 70–9.

Ferguson, J. 1999. *Expectations of modernity: Myths and meanings of urban life on the Zambian Copperbelt.* Berkeley, CA: University of California Press.

Freudenburg, W.R. 1992. Addictive economies: Extractive industries and vulnerable localities in a changing world economy. *Rural Sociology* 57: 305–32.

Freudenberg, W.R., and L.J. Wilson. 2002. Mining the data: Analyzing the socioeconomic effects of mining on rural communities. *Sociological Inquiry* 72: 305–22.

Freund, W.M. 1981. Labour migration to the Northern Nigerian tin mines, 1903–1945. *Journal of African History* 22, no. 1: 73–84.

Freund, W. 2007. *The African city: A history.* Cambridge: Cambridge University Press.

Gluckman, M. 1961. Anthropological problems arising from African industrial revolution. In *Social change in modern Africa*, ed. A. Southall, 67–83. Oxford: Oxford University Press.

Godfrey, B.J. 1992. Migration to the gold-mining frontier in Brazilian Amazonia. *Geographical Review* 82, no. 4: 458–69.

Grätz, T. 2009. Moralities, risk and rules in West African artisanal gold mining communities: A case study of Northern Benin. *Resources Policy* 34: 12–7.

Harries, P. 1994. *Work, culture, and identity: Migrant laborers in Mozambique and South Africa, c. 1860–1910.* London: James Currey.

Hilson, G., and C. Potter. 2005. Structural adjustment and subsistence industry: Artisanal gold mining in Ghana. *Development and Change* 36, no. 1: 103–31.

Hilson, G., and N. Yakovleva. 2007. Strained relations: A critical analysis of the mining conflict in Prestea, Ghana. *Political Geography* 26, no. 1: 98–119.

Hilson, G., and A. Ackah-Baidoo. 2011. Can microcredit services alleviate hardship in African small-scale mining communities? *World Development* 39, no. 7: 1191–203.

Holliday, J.S. 2002. *The world rushed in: The California gold rush experience.* Norman, OK: University of Oklahoma Press.

Jeeves, A. 1985. *Migrant labour in South Africa's mining economy: The struggle for the gold mines'labour supply, 1890–1920.* Johannesburg: Witwatersrand University Press.

Johnston, R.J. 1982. *The American urban system: A geographical perspective.* Harlow: Longman.

Jønsson, J.B., and D.F. Bryceson. 2009. Rushing for gold: Mobility and small-scale mining in East Africa. *Development and Change* 40, no. 2: 249–79.

Kirk, J.F. 1991. Race, class, liberalism, and segregation: The 1883 Native Strangers' Local Bill in Port Elizabeth, South Africa. *The International Journal of African Historical Studies* 24, no. 2: 293–321.

Kohrs, E.V. 1974. Social consequences of boom growth in Wyoming. Paper presented at the Rocky Mountain American Association of the Advancement of Science Meeting, April 24–26, Laramie, Wyoming.

Laverty, J.R. 1983. Urban development. In *Mining and Australia*, ed. W.H. Richmond and P.C. Sharma, 119–49. St. Lucia: University of Queensland Press.

Lawrie, M., M. Tonts, and P. Plummer. 2011. Boomtowns, resource dependence and socio-economic well-being. *Australian Geographer* 42, no. 2: 139–64.

Le Billon, P. 2001. The political ecology of war: Natural resources and armed conflicts. *Political Geography* 20: 561–84.

Le Billon, P., and E. Levin. 2009. Building peace with conflict diamonds? Merging security and development in Sierra Leone. *Development and Change* 40, no. 4: 693–715.

Luning, S.W.J. 2006. Artisanal gold mining in Burkina Faso: Permits, poverty and perceptions of the poor in Sanmatenga, the land of gold. In *Small-scale mining, rural subsistence and poverty in West-Africa*, ed. G.M. Hilson, 135–48. Rugby, UK: Practical Action Publishing.

Mabin, A. 1986. Labour, capital, class struggle, and the origins of residential segregation in Kimberly, 1880–1920. *Journal of Historical Geography* 12, no. 1: 4–26.

Mabin, A. 1989. Struggle for the city: Urbanisation and political strategies of the South African state. *Social Dynamics* 15, no. 1: 1–28.

Mkandawire, T. 2002. The terrible toll of post-colonial 'rebel movements' in Africa: Towards an explanation of the violence against the peasantry. *Journal of Modern Africa Studies* 40, no. 2: 181–215.

Moodie, T.D. 1994. *Going for gold: Men, mines and migration.* Berkeley: University of California Press.

Morris, R.J. 1986. Urbanisation. In *Atlas of industrializing Britain 1780–1914*, ed. J. Langton and R.J. Morris, 164–79. London: Methuen and Co. Ltd.

Morris, R.J. 1993. Urbanisation. In *The Victorian city: A reader in British urban history*, ed. R.J. Morris and R. Rodger, 43–72. Longman: Harlow.

O'Connor, A. 1983. *The African city.* London: Hutchinson & Co.

Parks, R. 2011. 'Can I go?' – Exiting the artisanal mining sector in the Democratic Republic of Congo. *Journal of International Development* 23: 1115–27.

Peters, K., and P. Richards. 1998. 'Why we fight': Voices of youth combatants in Sierra Leone. *Africa* 68, no. 2: 183–210.

Potts, D. 2010. *Circular migration in Zimbabwe and contemporary Sub-Saharan Africa.* Woodbridge UK: James Currey and New York: Boydell & Brewer Inc.

Rajak, D. 2012. Platinum city and the New South African dream. *Africa* 82, no. 2: 252–71.

Rakodi, C. 1986. Colonial urban policy and planning in Northern Rhodesia and its legacy. *Third World Planning Review* 8, no. 3: 193–217.

Richards, P. 1996. *Fighting for the rain forest: War, youth and resources in Sierra Leone.* London: James Currey.

Richards, P. 1999. Youth war in Sierra Leone: Pacifying a monster? Paper presented at the 11th International Colloquium on Ethnic Construction and Political Violence, Cortona Centros S. Agostino, July 2–3, 1999.

Ross, M. 2001. *Extractive sectors and the poor: An Oxfam America report.* Boston, MA: Oxfam America.

Saad, E.N. 1983. *Social history of Timbuktu: The role of Muslim scholars and notables, 1400–1900.* Cambridge: Cambridge University Press.

Shankleman, J. 2011. The impact of mining on urbanisation: A large-scale mining perspective. Paper presented at the Conference on the Impact of Mining on Urbanisation, UPIMA, Bagamoyo, Tanzania, February 27, 2011.

Smith, M.D., R. Krannich, and L.M. Hunter. 2001. Growth, decline, stability and disruption: A longitudinal analysis of social well-being in four western rural communities. *Rural Sociology* 66, no. 3: 425–50.

Storey, K. 2001. Fly-in/fly-out and fly-over: Mining and regional development in Western Australia. *Australian Geographer* 32, no. 2: 133–48.

Swanson, M. 1977. The sanitation syndrome: Bubonic plague and urban native policy in the Cape Colony, 1900–1909. *Journal of African History* 18, no. 3: 387–410.

Tonts, M., and M. Taylor. 2010. Location patterns of large corporate headquarters in the Australian urban system. *Urban Studies* 47: 2641–64.

UNCTAD (United Nations Conference on Trade and Development). 2007. *World investment report 2007, transnational corporations, extractive industries and development.* New York and Geneva: United Nations.

United Nations Population Division 2004. *World urbanisation prospects: The 2003 revisions.* New York: Economic and Social Affairs, Population Division.

United Nations MONUC (Mission in the DRC) 2004. *Principal cities, Congo (Dem Rep.),* Department of Peace Keeping Organisation GIS Unit. http://www.citypopulation.de/CongoDemRep.html (accessed April 22, 2012).

Vennetier, P. 1976. *Les villes de l'Afrique Noire.* Paris: Masson.

Verhoeve, A. 2004. Conflict and the urban space: The socio-economic impact of conflict on the city of Goma. In *Conflict and social transformation in Eastern DR Congo,* ed. K. Vlassenroot and T. Raeymaekers, 103–22. Gent: Academia Press Scientific Publishers.

Vlassenroot, K., and K. Büscher 2009. The city as frontier: Urban development and identity processes in Goma. Working Paper no. 61, London School of Economics, Crisis States Research Centre.

Walker, R.A. 2001. California's golden road to riches: Natural resources and regional capitalism, 1848–1940. *Annals of the Association of American Geographers* 91, no. 1: 167–99.

Walsh, A. 2003. 'Hot money' and daring consumption in a northern Malagasy sapphire-mining town. *American Ethnologist* 30, no. 2: 290–305.

Walsh, A. 2012. After the rush: Living with uncertainty in a Malagasy mining town. *Africa* 82, no. 2: 235–51.

Welsby, D.A. 1996. *Kingdom of Kush: The Napatan and Merotic empires.* London: British Museum Press.

Werthman, K. 2009. Working in a boom-town: Female perspectives on gold-mining in Burkina Faso. *Resources Policy* 34: 18–23.

Wilks, I. 1993. *Forests of Gold: Essays on the Akan and the Asante kingdom.* Athens USA: Ohio University Press.

Wilson, L.J. 2004. Riding the resource roller coaster: Understanding socioeconomic differences between mining communities. *Rural Sociology* 69, no. 2: 261–81.

Wolpe, H. 1972. Capitalism and cheap labour power in South Africa: From segregation to apartheid. *Economy and Society* 1, no. 4: 425–54.

Mining, housing and welfare in South Africa and Zambia: an historical perspective

Hugh Macmillan

This contribution seeks to place the relationship between mining, housing and welfare in historical perspective in two Southern African countries with long histories of mining activity that continue to have large mining sectors, namely South Africa and Zambia. Mining capitalists and colonial, as well as apartheid, governments tended to resist urban permanence, but they were usually unable to stop it. Economists and social anthropologists, being less conscious than historians of the cyclical nature of mining, have been prone to discount the remarkable tenacity of the Zambian Copperbelt's urban residents in the face of mining downturns. By contrast, very little attention in the literature has been paid to the decline of gold mining in South Africa. In view of the Zambian precedent for urban resilience, it would be unwise to overlook the continuing influence of South African mining on urban settlement in spite of the industry's gradual decline over the last four decades. The patterns of development of new platinum mines in South Africa and new copper mines in Zambia show certain similarities and historical continuities, including a reluctance on the part of mining companies to accept responsibility for the housing and welfare of all their workers and their families.

'Camp', 'compound', 'township' – these specialised terms for housing settlements associated with mining in Southern Africa over the last century and a half suggest that there is something different about mines and urbanisation in the region. 'Camp' implies something temporary and impermanent; 'compound' is a loan word from Malay – *kampong* – and originally referred to an enclosure, or to an ethnic zone in a village; 'township' is an old English word, but it seems to have acquired a peculiar Southern African resonance, as not just a small town, but something less than a town, one lacking in the usual urban facilities.

Of course, mining is a peculiar industry. Mines are sited as a result of geological accidents and may be far removed from fresh water, good soils, the coast, or the navigable rivers that usually contribute to the organic growth of towns. Mine towns are always new towns and they may grow up in places that would not normally support a large population – think of the Witwatersrand, the Copperbelt, or in Botswana, Selebi Pikwe or Orapa (Gwebu in this collection), or in Namibia, Tsumeb. Mines take a long time to come into production and may then have short lives. They often employ more people in the construction phase than they do in the production phase, and established mine towns, even company towns, tend to provide more employment in services, formal or informal, than they do in actual mining.

Mining is the most cyclical of industries and investment in it is often speculative. Commodity markets are characterised by a pattern of increased demand, shortages, high prices, new investment, increased production and over-supply – of boom and bust. Historically, mines needed to pay workers, taxes and royalties, repay capital investment, provide a return to investors and generate funds for development, within a short time-frame. In order to do this, they required small quantities of expensive skilled labour, large quantities of cheap, unskilled labour and high rates of profit.

The initial South African solution to these problems is extremely well known: the employment of small numbers of skilled and semi-skilled white workers, and of large numbers of cheap black workers through the development of a system of oscillating labour migration between rural 'reserves' and urban 'compounds'. Initially voluntary, the flow of labour was expanded and maintained through taxation, the provision of credit and a system of organised labour recruitment that extended far beyond the borders of South Africa itself – as far as southern Tanzania (Jeeves 1985; Wilson 1972; Lacey 1981; Crush et al. 1991).

This was, from a Marxist point of view, and possibly a Weberian one, a system of semi-proletarianisation with the reproduction and retirement of the bulk of the workforce funded through the labour of older men, women and children in rural areas. It was a system of exploitation with the profits of mining capitalists subsidised by rural families. It was a system that actively discouraged urbanisation, at least for black people, who were seen as 'temporary sojourners' in town. Permanent urbanisation was not officially on the agenda. If mining 'camps', such as Kimberley or Johannesburg, developed into towns, with substantial and increasingly permanent black as well as white populations, this was because they developed as financial, commercial, manufacturing and service centres, and it was in spite of mining capitalists, governments, pass laws, which were systematised on a national basis in 1923 and hugely reinforced by the apartheid government in 1952, and forced removals, which were a regular feature of urban life for Africans from the 1950s to the 1980s (Delius 1983; Worger 1987; Turrell 1987; Chipkin 1993).

The housing of Africans in Johannesburg and the Witwatersrand

Where systems of labour migration were in place, as in Johannesburg and the Witwatersrand, it is axiomatic that the permanent or semi-permanent African population emerged from among unrecruited workers who were employed initially in domestic, commercial and other service industries, and, in large numbers from the Second World War onwards, in manufacturing. There was always, however, some leakage from the recruited migrant labour force into the settled urban population. In 1919 it was estimated that half the overwhelmingly male African population of 100,000 in Johannesburg was employed in mining, but this proportion fell steadily with the decline of mining, the growth of other sectors, and an increasing population of women and children. Gender balance in the African population of Johannesburg, implying a measure of stability, was reached in 1967. Some official recognition of urban permanence had come with the building by the Johannesburg municipality of Orlando Township to the south-west of the city in 1931. This was the nucleus of what was to grow into the complex of townships, which became known, though not until 1963, as Soweto (Bonner and Segal 1998).

The real turning point for the African population of Johannesburg was the Second World War when, as a result of import substitution, the development of manufacturing and the withdrawal of a significant proportion of the white male population for the war effort, there was a hugely increased demand for labour, with higher wages paid than in mining or agriculture. There was a temporary suspension of the pass laws, and the influx was initially housed by the infilling of shacks in existing urban settlements and then by the occupation of vacant land by squatters. There was no provision of new housing by the city or the state until after the war. Following the election of the National Party government in 1948, and the gradual evolution of the policy of apartheid, a massive programme of housing construction was begun in the early 1950s by the Johannesburg City Council, with some loan funding from the mines, and by the state. From the point of view of one of the leading ideologists of apartheid, W.M.M. Eiselen, as stated in 1951, the purpose of this programme was not philanthropy, but control.

> Only by the provision of adequate shelter in properly planned Native townships can full control over urban Natives be regained, because only then will it be possible to eliminate the surplus Natives who do not seek or find an honest living in the cities. (Chipkin 1993, 211)

It is easier to control people living in formal than in informal settlements. By the late 1960s, over 100,000 more or less identical two-to-three bedroom 'matchbox' houses had been built over an area of 150 square kilometres to the south-west of Johannesburg, with a population of about 1 million people. This has been accurately described as a 'dormitory area on the edge of town – a sterile, anti-social, polluted, urban environment devoid of infrastructure and facilities' (Chipkin 1993, 216). By the early 1970s the building of family homes had ceased, and migrant hostels had been planted within the townships, as part of a latter-day attempt to impose migrancy from the so-called 'homelands' on some non-mining employees in Johannesburg.

The introduction of labour migrants into a settled population with much-prized 'section ten' rights to permanent residence in Johannesburg, part of a policy of divide-and-rule, was to have explosive consequences. Among the few facilities that were provided by the state in Soweto were schools, and it was from them, and the alienated youth, that the Soweto Uprising erupted in 1976. In spite of massive repression over a decade, the state was never able to regain control over Soweto or stop the influx from the rural areas to town. The repeal of the pass laws in 1986 was a belated recognition of political and administrative failure. A new growth of shacks had appeared in Soweto by the early 1980s, and by the end of the decade there were an estimated 42,000 such temporary back-yard dwellings within the townships (Bonner and Segal 1998).

The late 1980s also saw the beginning of the breakdown of the racial zoning of inner-city Johannesburg, which had been imposed by the Group Areas Act, and a new African influx into areas such as Hillbrow and Yeoville. This was followed by a post-apartheid, and 'cosmopolitan', influx of people from independent Africa, from Zimbabwe to the Congo and Senegal, many of whom eked out a living as petty traders and hawkers in a burgeoning informal economy. The appearance of visible poverty and street crime in Johannesburg's inner city provoked the flight of businesses and better-off residents to the northern suburbs, a process that has been partially reversed in recent years with a new process of 'gentrification' and the conversion to upmarket flats of disused office accommodation.

Debates about segregation and apartheid

There has, of course, been huge debate as to whether the political systems of segregation and later apartheid were developed in the interests of mining capital, agricultural capital, or of white labour, or a combination of some of these (Lipton 1985, 2007). There are no simple answers. Historians can show that the pass laws, the pattern of so-called 'Native reserves', punitive Masters and Servants legislation and labour migration all predated the mining revolution of the late nineteenth century, and were not conceived by mining capitalists. The mine compound may have been an original contribution, though even it resembles the earlier slave lodge (Shell 1994). The industrial and mining colour bars that spread as far north as Zambia in the first half of the twentieth century were prompted by the reaction of white workers to the attempts of mine-owners to reduce the number of white workers and to replace them with cheaper black workers. The relative cheapness of black labour stemmed in part from the fact that black workers were only semi-proletarianised and had rural homes to which they could return. This was increasingly impossible for wholly proletarianised white workers of the landless squatter or *bywoner* class, some of whom were driven into shanty towns and squatter camps around Johannesburg by the Anglo-Boer War, while others were gradually extruded from the rural areas in the following decades (Krikler 2005; Wilson 1972).

The profits of mining capitalists depended on the system of labour migration, as was acknowledged during the Second World War by William Gemmill, the architect of the Witwatersrand Native Labour Association (WENELA) and of South Africa's sub-regional 'labour empire' in evidence to the Lansdown Commission in 1943. But mining capitalists did not necessarily welcome apartheid and sometimes funded opposition parties, as well as urban housing. Mining capitalists, including, for example, the Oppenheimers, generally disliked industrial colour bars or any legislation that restricted their ability to hire and fire workers, or to substitute cheaper black workers for expensive white ones. They were not opposed in principle to the emergence of a black labour aristocracy and campaigned unsuccessfully for permission to build housing for at least a small percentage of the black workforce on the post-war Free State gold mines. A black labour aristocracy, and a measure of urban stabilisation, did occur in the Congo copper mines and eventually in Northern Rhodesia/Zambia, but never, for mine labour anyway, in South Africa (Gregory 1962; Innes 1984; Perrings 1979).

Oscillating labour migration and stabilisation in central Africa

The Great Depression caused a crisis of rural impoverishment in South Africa in the 1930s and there were commissions of enquiry into both black and white poverty. There was contemporary alarm about the impact of labour migration – especially the withdrawal of young men – on rural areas there and elsewhere, but social anthropologists and historians working in northern Zambia, Pondoland and southern Mozambique have shown since the 1950s that oscillating labour migration did not necessarily impoverish rural areas. It often provided necessary capital for agricultural investment and contributed not so much to 'detribalisation' as to 'tribal cohesion' (Watson 1958; Beinart 1982; Harries 1994).

These arguments originated in the work of the Cambridge economist E.A.G. (Austin) Robinson, a right-hand man to J.M. Keynes. The development of the Zambia/Congo Copperbelt from the late 1920s into the second great mining area of the region had prompted the first systematic study by a group of social scientists of the future impact of mining in underdeveloped areas. J. Merle Davis's edited volume, *Modern Industry and the African* (1933), and the chapter in it by Robinson entitled 'The Economic Problem', was hugely influential, and has been seen as a foundation text in development economics. Writing at the low point of the Great Depression, and before the new mines came on stream, Robinson recognised that a degree of urbanisation was inevitable, but he did not think that it was necessarily desirable. He was acutely conscious of the cyclical nature of mining and the uncertain lifespan of mines. Writing as a Keynesian liberal, he put a positive spin on the Marxist interpretation of labour migration. He doubted the wisdom, or morality, of creating an industrial proletariat and a permanently urbanised population in a mining area where there was no social welfare provision for future unemployment in cyclical downturns or for ultimate retirement. He worried about breaking the links between the urban and the rural areas, about the development of a dual economy, and he thought that oscillating labour migration was a lesser evil (Robinson 1933; Macmillan 1996).

Although there was a movement towards the stabilisation of mine labour on the Congolese Copperbelt from the 1930s, it was not until after the Second World War that the government of Northern Rhodesia and the mines began to make large-scale provision for family housing. They were then motivated by the need to create a stable and more highly skilled workforce, but the housing was only for people in current employment. Not all urban housing was built by the mines. Much of it was built by municipalities for the housing of other service workers, and all large employers were required to provide housing for their workers. By the late 1960s Zambia's copper miners were the best-paid workers in black Africa, and by the early 1970s Zambia was the most urbanised country in Africa outside of South Africa. The urban population grew rapidly after independence in 1964 as a result of higher wages, the increased stabilisation of mine labour, the abandonment of controls on settlement, a more permissive attitude to squatters and schemes to upgrade squatter settlements, and, later on, by food subsidies and other policies that encouraged rural exodus. The urban population tripled between 1964 and 1980 as a proportion of the whole, though much of the increase was in the capital, Lusaka, and it was not all directly dependent on mining (Ferguson 1990a, 1990b; Macmillan 1993).

With the decline of the Copperbelt mining industry as a result of the long depression in base metal prices that began in the mid-1970s, a decline that was accentuated by nationalisation, deeper-level mining, lower ore grades, a failure to reinvest in mine development and increased transport costs, some saw Robinson's nightmare scenario of the emergence of an unemployed urban proletariat as coming to pass. Employment in copper-mining peaked in the mid-1970s at around 56,000 and, as a result of an unspoken compact between the government and the mineworkers' union, remained at a stable, but uneconomic, level until the early 1990s. Real incomes fell dramatically and relative impoverishment and deteriorating living standards coincided with the onset of HIV/AIDS and falling life expectancy from the mid-1980s. Copper production fell gradually from about 760,000 tons per annum in 1969 to 257,000 tons in 2000 and the copper price reached a low point of $1300 a ton in the following year (ILO 1977; Chamber of Mines 2011).

Doing fieldwork on the Copperbelt in 1985–1986, the American social anthropologist, James Ferguson, took issue with what he described as 'the modernist narrative', which, he alleged, affected the social anthropologists of the influential Rhodes–Livingstone Institute/Manchester School. He argued that they thought in terms of an inevitable transition to urban permanence, which they identified with an industrial revolution and 'modernisation'. He doubted that urban permanence ever happened on the Zambian Copperbelt, preferring to see a continuously shifting and transitory urban population. At the same time he had little doubt that the malaise affecting the Zambian Copperbelt in the 1980s and 1990s was terminal. 'Expectations of modernity' and of an industrial revolution had been crushed. His apocalyptic, or more precisely eschatological, vision of the 'last days' of the Copperbelt was deeply influenced by the time, now nearly 30 years ago, and the place, Kitwe, at which he worked, in what were in reality the last days, not of the Copperbelt, but of what was by then the deeply unpopular regime of President Kaunda and the United National Independence Party (UNIP). They had presided over 15 years of cyclical, though not terminal, decline and were soon to be removed from office after two botched attempts, in 1986 and 1990, to implement necessary financial reforms including the reduction of urban food subsidies (Ferguson 1990a, 1990b, 1994, 1999; and author's recollections).

Writing in response, I doubted that there ever was a 'modernist narrative' because writers on the Copperbelt were almost always painfully aware of the cyclical nature of the mining industry. I did, however, suggest that census statistics showed increasingly equal numbers of males and females on the Copperbelt – something that was normally seen as evidence of a degree of permanence. I also suggested that the Copperbelt might not be a victim of terminal decline, but rather of an unusually long cycle. I was also sceptical about levels of reverse migration to remote rural areas and about what Deborah Potts has more recently described as counter-urbanisation or de-urbanisation (Potts 2005, 2010). Clearly there was a small amount of reverse migration to rural areas, but movement from the Copperbelt to other towns such as Lusaka and increased levels of squatting on mine land and in forest reserves within the Copperbelt were more pronounced. Some Copperbelt urban residents sought to make a living through market gardening and charcoal burning in peri-urban areas within the province. Although the population growth rates for Copperbelt towns fell below the national average from the 1980s onwards, and the level of urbanisation in Zambia as a whole remained static, there was little net reduction in the urban population of the Copperbelt towns.

Coping strategies and partial recovery

Not much attention has been paid to the remarkable tenacity of the Copperbelt's urban residents in the face of real adversity, except by Hansungule, Feeney and Palmer in their Oxfam survey of the plight of squatters on Zambia Consolidated

Copper Mines (ZCCM) land in 1998 during the run-up to the delayed and mismanaged privatisation of the mines, Patience Mususa's more recent work in this collection on the 'greening' of Luanshya, the most devastated of the Copperbelt towns and the victim of the most predatory and cynical of the new private owners. Both these studies point, however, to a blurring of the boundaries between the urban and the rural on the Copperbelt, something that may also be the case in the new mining areas in the North West and around Lusaka (Macmillan 1993, 1996; Hansungule et al. 1998; Gewald and Soeters 2010).

Copper prices did eventually recover, though not for 30 years. Recovery began in 2004, and copper prices have, with some extreme fluctuations in the immediate aftermath of the credit crunch in 2007–2008, continued to rise, reaching $10,000 a tonne in February 2011, declining to a still highly viable $7500 a tonne in May 2012. Zambian copper production in 2010 exceeded 800,000 tonnes, breaking the record of 760,000 tonnes achieved in 1969, the year before nationalisation. The record was broken again in 2011, with output around 850,000 tonnes. Employment in mining in 2009 had fallen from a 2007 peak of 38,000 to 32,000 as a result of a price crash in late 2008 and probably also as a result of a slow-down in mine construction work. Employment levels may now be close to 40,000, not far below the level at independence in 1964, though the population of Zambia has tripled since that date (Chamber of Mines 2010; Zambia 2011).

With new investment of $5 billion in the last decade, some of which has already been rewarded through exceptionally generous development agreements and tax breaks, and high profits during the 2004–2007 boom, there has been a partial recovery of mining employment on the old Copperbelt. The old mines are all working, with the possible exception of Luanshya where the new Chinese owners – the third owners since privatisation – have begun a process of refurbishment. There is also a new and highly productive, though high-cost, deep-level mine at Konkola near Kitwe controlled by the Indian mining company, Vedanta. The town of Kitwe itself is said to be booming, with traffic jams, new private housing, new shopping malls – one of which is being constructed on the site of the splendid Nkana Hotel, an iconic symbol of the old Copperbelt. In Ndola, always primarily an administrative and commercial rather than a mining centre, the cement factory is back in business. Even Zambia's oldest major mine, the lead-zinc mine at Kabwe, the scene of Godfrey Wilson's research on 'the economics of detribalisation', about 100 miles south of the Copperbelt, which closed in 1996, is being re-opened for the processing of the mine dumps and tailings and is expected to employ 1000 workers (Chamber of Mines of Zambia website, accessed 23 May 2012; information from Murray Sanderson, Kitwe).

The centre of gravity for mining is, however, shifting to new, or expanded, open-cast mines at Kansanshi and Lumwana on an extension of the Copperbelt around Solwezi about 150 miles, and further, to the west. As an example of this kind of development, First Quantum's Lumwana mine is designed to produce 125,000 tonnes of copper a year for 40 years. There are plans for a further three large open-cast mines in this area. As Patience Mususa points out in her piece in this collection, the paternalistic welfare capitalism, which was imposed on the old mining companies by the colonial state, and maintained by their parastatal successors, has given way to a rougher and tougher brand. Mechanisation, sub-contracting and out-sourcing have weakened the links between mining employers and employees. Mine townships were handed over to the municipalities in the 1990s, mine housing was sold off during the process of privatisation, and health, education and welfare provision has been either eliminated or reduced. By comparison with the old Copperbelt mines, the new mining developments employ relatively few workers, but formal housing is only provided for the skilled minority of that smaller workforce. Squatter camps for general workers, informal service workers and work-seekers have grown up around the new mining developments in the North West.

Arguments about 'alternative modernities', 'disconnection and abjection', counter-urbanisation, de-urbanisation, reverse and circular migration may have more relevance in Zimbabwe where mining centres were relatively small and scattered, and where there was more of a balance in the economy between mining, manufacturing and commercial agriculture, than they do on the Copperbelt, or the Witwatersrand (Potts 2010). Amin Kamete's fascinating paper in this collection demonstrates an extraordinary and, it is to be hoped, exceptional recent history of 'ghost towns and havens', forced removals from farms and towns, and enforced repatriation of foreign workers.

Mining and urban housing in the new South Africa

In South Africa, too, there have been dramatic changes since the 1980s. The end of the pass laws allowed mine workers to take their families to town, but the mines were under no obligation to house them. Families often lived in new squatter 'camps'. Increased wages and better conditions of employment prompted mechanisation and sub-contracting.

The balance of employment in Johannesburg had by then changed once again. From 1980, declining employment in manufacturing marked a process of de-industrialisation. By 1990 there were five times as many people employed in services, including commerce and finance, as there were in the mining and manufacturing sectors combined (Beall et al. 2002). Mining now makes only a small contribution to the economy of Johannesburg, and the surrounding province of Gauteng, although the city has not cut its links with mining and continues to be a major financial, commercial and service centre for mines throughout the region.

While a great deal of academic attention has been devoted to the allegedly terminal, though actually cyclical, decline of the Zambian copper industry, remarkably little attention has been devoted to the apparently terminal decline of the South African gold mines and the rise of platinum mining (Nattrass 1995; Crush and James 2002). Gold production in South Africa reached 1000 tons in 1970, constituting more than two-thirds of world production. It has since fallen by 80% despite astonishingly high gold prices on the world market, peaking at close to $2000 an ounce in 2011. No price response could be mustered from the main Free State's nearly exhausted mines, which were plagued by higher costs as a result of deeper levels and lower ore grades. South Africa now ranks fifth among the world's gold producers with a mere 7% of world production (www.goldsheetlinks.com/production.htm, accessed 25 May 2012).

By 2009, platinum had overtaken gold both in terms of its contribution to South Africa's GDP (2.4% to 2%) and to mining employment (184,000 to 160,000). In the early years of the millennium, the Anglo American Corporation's subsidiary, AngloGold, sold the whole of its Free State interests to Black Empowerment companies. AngloGold (now AngloGold Ashanti) was in the process of shifting its focus from South Africa to the rest of the world.

This decline, and the consequent reduction in employment, has clearly had a huge impact on the province. A recent newspaper report refers to the loss of tens of thousands of jobs and suggests that the rate of unemployment in the major Free State mining town of Welkom is 60% – twice the South African average and a higher rate than in most of Zambia's Copperbelt towns. The report also notes by way of colourful illustration that membership of the local Oppenheimer Park Golf Club has fallen from 1300 in the 1980s to 300 in 2012, while four of the holes are submerged beneath a polluted lake (England 2012). It is mysterious how such dramatic changes in the nature of South Africa's gold-mining industry, and in patterns of employment in an area like the Free State, have attracted so little academic, or journalistic, attention.

There may be a variety of reasons, including the diversity and complexity of the South African economy, but also, perhaps, the fact that as old gold mines closed, or reduced their output, new, and more highly mechanised, platinum, coal and iron ore mines took their place. The most spectacular growth has been in platinum production in the Rustenburg district, most of it on land belonging to the Royal Bafokeng Authority, a Tswana traditional authority. Rustenburg became the fastest-growing town in South Africa: the population of the municipality doubled from 250,000 to 500,000 between 1995 and 2010 and it is expected to double again in the next 15 years.

It is estimated that more than one-third of Rustenburg's housing units are informal – in other words, shacks – and almost one-third of the population lives in them. Although new 'housing suburbs' have been built for managerial and skilled workers, many of them white, and accommodation for some of the unskilled and semi-skilled workers has been provided in single-sex hostels, a significant proportion of the mineworkers employed live in informal settlements. Some workers have rooms in hostels and keep their families in squatter camps. Some hostels have been converted for family accommodation, but many workers use a housing allowance to supplement low wages and live with their families in the camps. Schemes for the upgrading and formalisation of some squatter settlements have been resisted by the Royal Bafokeng Authority, which is reluctant to allow the settlement of 'migrants' on its land, possibly fearing that such people might acquire a claim to a share in the income derived by the 'tribe' from the exploitation of platinum. There is a tendency for the mining companies as well as the authority to see 'informal settlements as transitory, impermanent, and usually illegal' and to categorise their 'migrant' inhabitants as rightless 'non-stakeholders' (Rajak 2012; and communication from Professor Bernard Mbenga, University of the North West, 2012). Issues relating to wages and housing, as well as competition between the National Union of Mineworkers and a breakaway union, contributed to the strikes and occupation during which police shot and killed 34 workers at Lonmin's Marikana Mine in August 2012.

The number of people employed on the mines in South Africa as a whole has fallen from about 720,000 in 1984 to about 440,000 in 2009 (Race Relations Survey 1985, 220; South Africa Survey 2011, 336). The latest statistics from the Chamber of Mines, which counts employees in the 'mining sector', rather than on the mines themselves, record a higher figure of 500,000 for 2011 (Chamber of Mines of South

Africa 2011). Organised recruitment from within South Africa and outside of it is no longer necessary and the flow of remittances to the rural areas has slowed down and been replaced within South Africa by state pensions. There are still some foreign workers on South Africa's mines, but the number has been greatly reduced, and there has been no replacement for remittances to the traditional labour recruitment areas outside South Africa.

South Africa is probably different from every other African state, and better able to withstand the impact of the decline of mines in some areas, as it has almost universal social welfare provision – no unemployment benefit, but disability and old age pensions, the value of which has, perhaps surprisingly, kept pace with inflation over the last 20 years (South Africa Survey 2011). With the addition of single-parent families to the welfare system, the number of beneficiaries has greatly increased over the past decade. South Africa also has liberal labour laws and a minimum wage, though some argue that these are a mixed blessing, contributing to a declining proportion of people in formal employment in all sectors, not just mining, but especially agriculture, since 1994 (South Africa Survey 2011, 226, 416).

Conclusion

The relationship between mining, urbanisation and poverty has a great deal to do, as Austin Robinson pointed out in 1933, with trade cycles, the life-span of mines and social welfare provision, or its absence. Oscillating labour migration was the negation of urbanisation and could be seen, on the one hand, as a system that exploited and impoverished rural areas, or, on the other hand, as a system that maintained urban–rural links, provided capital for agricultural improvement and was a substitute for state-sponsored social welfare mechanisms. Mining capitalists and colonial, as well as apartheid, governments tended to resist urban permanence, but they were often unable to stop it. The question of what happens *after mining* is not one that is confined to Africa, but also relates, for example, to the formerly coal-mining communities of the United Kingdom and the United States, but what happens to ex-mining communities without social welfare mechanisms and safety nets is more of an African or a developing world question.

There is one other concept that may be relevant to the study of mining, urbanisation and poverty, and that is the notion of the 'resource curse' (Auty 1993). This is the idea that discoveries of minerals, whether gold, diamonds, platinum, copper or oil, may have seriously distorting effects on economies. Their exploitation may provide revenues that are easily diverted by kleptocracies, or that are too easy to collect, discouraging governments from promoting genuinely mixed economies and enlarged tax bases. Minerals may provide easy sources of foreign exchange, encourage over-valued currencies and cheap imports, and so discourage local agriculture and manufacturing. Minerals are wasting assets and their exhaustion may leave huge social and environmental problems, including urban unemployment and rural impoverishment, as well as toxic dumps and polluted water supplies. All too often the governments, and the corporations that benefit from their exploitation, fail to make compensatory social welfare provisions or to invest the profits locally.

This article has focused on two countries in Southern Africa – South Africa and Zambia – which both have long histories of mining activity and continue to have large mining sectors. Economists and social anthropologists have frequently written off the Copperbelt, being less conscious than historians of the cyclical nature of mining. Very little attention appears to have been paid to the decline of gold mining in South Africa, but on the Zambian precedent it would be unwise to write off the industry in spite of the appearance over the last 30 or 40 years, most markedly in the last decade, of decline. Higher prices and new technologies may well bring new and longer life to apparently moribund mines and their associated towns. The patterns of development of new platinum mines in South Africa and of new copper mines in Zambia show certain similarities, and continuities with the past, including a reluctance on the part of mining companies to accept responsibility for the housing and welfare of all their workers and their families.

Notes on contributor

Hugh Macmillan is an historian who has taught at universities in Swaziland, Zambia and South Africa. His most recent book as author is *An African Trading Empire* (2005) and as editor *Mona's Story* (2008). He is currently completing a book on the history of the African National Congress (of South Africa) in exile in Zambia and is a research associate of the African Studies Centre at Oxford University. He can be contacted at: hughmacm@gmail.com

References

Auty, R.M. 1993. *Sustaining development in mineral economies: The resource curse thesis.* London: Routledge.

Beall, J., O. Crankshaw, and S. Parnell. 2002. *Uniting a divided city: Governance and social exclusion in Johannesburg.* London: Earthscan.

Beinart, W. 1982. *The political economy of Pondoland, 1860 to 1930.* Johannesburg: Ravan Press.

Bonner, P., and L. Segal. 1998. *Soweto, a city.* Cape Town: Maskew Miller, Longman.

Chamber of Mines of Zambia. 2010. Chamber of Mines of Zambia Presentation to the Public Discussion Forum to be made on 16 July at Solwezi. Economics Association of Zambia website: http://www.eaz.org.zm.

Chamber of Mines of South Africa. 2011. Annual Report.

Chipkin, C.M. 1993. *Johannesburg style: Architecture and society, 1880s–1960s.* Cape Town: David Philip.

Crush, J., A. Jeeves, and D. Yudelman. 1991. *South Africa's labor empire: A history of black migrancy to the gold mines.* Boulder, CO: Westview Press.

Crush, J., and W. James. 2002. Depopulating the compounds: Migrant labor and mine housing in South Africa. *World Development* 19, no. 4: 301–16.

Davis, J.M. 1933. *Modern industry and the African.* London: International Missionary Council.

Delius, P. 1983. *The land belongs to us.* Johannesburg: Ravan Press.

England, A. 2012. South African gold town struggles to find a future. *Financial Times*, May 9.

Ferguson, J. 1990a. Mobile workers, modernist narratives: A critique of the historiography of transition on the Zambian Copperbelt, Part 1. *Journal of Southern African Studies* 16, no. 3: 385–412.

Ferguson, J. 1990b. Mobile workers, modernist narratives: A critique of the historiography of transition on the Zambian Copperbelt', Part 2. *Journal of Southern African Studies* 16, no. 4: 603–2.

Ferguson, J. 1994. Modernist narratives, conventional wisdoms, and colonial liberalism: Reply to a straw man. *Journal of Southern African Studies* 20, no. 4: 633–40.

Ferguson, J. 1999. *Expectations of modernity: Myths and meanings of urban life on the Zambian Copperbelt.* Berkeley: University of California Press.

Gewald, J.-B., and S. Soeters. 2010. African miners and shape-shifting capital flight: The case of Luanshya-Baluba. In *Zambia, mining and neoliberalism: Boom and bust on the globalized Copperbelt*, ed. A. Fraser and M. Larmer, 155–84. New York: Palgrave Macmillan.

Gregory, T. 1962. *Ernest Oppenheimer and the economic development of Southern Africa.* London: Oxford University Press.

Hansungule, M., P. Feeney, and R. Palmer. 1998. *Report on land insecurity on the Zambian Copperbelt.* Oxford: Oxfam GB.

Harries, P. 1994. *Work, culture and identity: Migrant laborers in Mozambique and South Africa, 1860–1910.* Johannesburg: Witswatersrand University Press.

Innes, D. 1984. *Anglo: Anglo American and the rise of modern South Africa.* Braamfontein: Ravan Press.

International Labour Office (ILO). 1977. *Narrowing the gaps: Planning for basic needs and productive employment in Zambia.* Addis Ababa: International Labour Office, Jobs and Skills Programme for Africa.

Jeeves, A. 1985. *Migrant labour in South Africa's mining economy, 1890–1920.* Kingston: McGill University Press.

Krikler, J. 2005. *The Rand revolt: The 1922 insurrection and racial killing in South Africa.* Johannesburg: Jonathan Ball.

Lacey, M. 1981. *Working for Boroko: The origins of a coercive labour system in South Africa.* Johannesburg: Ravan Press.

Lipton, M. 1985. *Capitalism and apartheid: South Africa, 1910–6.* Aldershot: Wildwood House.

Lipton, M. 2007. *Liberals, Marxists and capitalists: Competing interpretations of South African history.* London: Palgrave Macmillan.

Macmillan, H. 1993. The historiography of transition on the Zambian Copperbelt – Another view. *Journal of Southern African Studies* 19, no. 4: 681–712.

Macmillan, H. 1996. More thoughts on the historiography of transition on the Zambian Copperbelt. *Journal of Southern African Studies* 22, no. 2: 309–12.

Nattrass, N. 1995. The crisis in South African gold mining. *World Development* 23, no. 5: 857–68.

Perrings, C. 1979. *Black mineworkers in Central Africa.* London: Heinemann.

Potts, D. 2005. Counter-urbanisation on the Copperbelt: Interpretations and implications. *Urban Studies* 42, no. 4: 583–609.

Potts, D. 2010. *Circular migration in Zimbabwe and contemporary Sub-Saharan Africa.* Oxford: James Currey.

Race Relations Survey. 1985. *Race Relations Survey 1984.* Johannesburg: South African Institute of Race Relations.

Rajak, D. 2012. Platinum city and the new South African dream. *Africa* 82, no. 2: 252–71.

Robinson, E.A.G. 1933. The economic problem. In *Modern industry and the African*, ed. J.M. Davis, 136–74. London: International Missionary Council Davis.

Shell, R.C.H. 1994. *Children of bondage: A social history of the slave society at the Cape of Good Hope, 1652–1838.* Johannesburg: University of the Witwatersrand Press.

South Africa Survey. 2011. *South Africa survey, 2010–11.* Johannesburg: South African Institute of Race Relations.

Turrell, R. 1987. *Capital and labour on the Kimberley diamond fields.* Cambridge: Cambridge University Press.

Watson, W. 1958. *Tribal cohesion in a money economy: A study of the Mambwe people of Zambia.* Manchester: Manchester University Press.

Wilson, F. 1972. *Labour in the South African gold mines, 1911–69.* Cambridge: Cambridge University Press.

Worger, W. 1987. *Kimberley: South Africa's city of diamonds.* New Haven, CT: Princeton University Press.

Zambia. 2011. *2010 Census of population and housing: Preliminary report.* Lusaka: Central Statistical Office.

The power of mining: the fall of gold and rise of Johannesburg

Philip Harrison and Tanya Zack

The City of Johannesburg has developed through the entire life-cycle of the mining industry. In its early years, its development was tied to the varying, but generally upward, fortunes of the mining industry. During this time, gold mining in Johannesburg, and along the Witwatersrand, propelled the growth of South Africa's national economy into a phase of self-sustained development, and created an integrated labour market across southern Africa. It also played a key role in shaping the racial oligarchy that dominated South Africa until the fall of apartheid in the 1990s. However, gold was eventually to decline, first in the areas around Johannesburg, and then elsewhere. The growth of Johannesburg, however, continued and the urban economy became increasingly diversified and flexible. This growth seemed divorced from mining but was, in fact, deeply rooted in the history of mining. The mining industry played an intimate role in the development of the manufacturing sector and also in the emergence of financial services; which is currently the leading economic sector in Johannesburg. These economic changes are represented in continuous evolution of the spatial form of the city. Currently the physical legacy of mining is understood mainly in terms of its deleterious environmental consequences, including acid mine drainage, with the long and profound impact of mining on the patterning of urban growth largely forgotten.

Will the gold-reef last; and the big city and the hiving population that have grown up around it – will these also endure? Or can it be that the Reef is approaching exhaustion, and that all its correlative interests are doomed to extinction? (Chilvers 1948, 222).

For almost the first half of the city's existence, there was a deep anxiety amongst its residents that Johannesburg would collapse when the mines closed. The mines eventually did shut down but the power of mining in Johannesburg is such that it ignited the development of an economy that outlived the mines and continued to grow and flourish.[1]

This contribution focuses on the development of Johannesburg through the full life cycle of the gold mines, and beyond. It shows how Johannesburg – in common with a few other cities in the world such as Melbourne and San Francisco – transcended the boom–bust scenario of a minerals-based economy and evolved into a diverse and competitive agglomeration. It supports the argument of Davis

(1998), which challenged a conventional view that economies initially dependent on mining inevitably have substandard economic performance in later years.

The analysis focuses specifically on the Central Rand Goldfield, where large-scale, deep-level gold mining first began in South Africa, and where Johannesburg was proclaimed as a mining settlement in 1886. Central Rand is one of seven distinct gold fields in South Africa, and during its long history of productive activity has produced 15% of South Africa's total gold output. Its importance in terms of physical production has however declined progressively – from 80% of total gold output nationally in 1911 to 3% in 1980 and nearly zero currently[2] – but the economic weight of the urban agglomeration it spawned continues to grow.

The development of Central Rand, and then of outlying gold fields along the ridge of hills known as the Witwatersrand, did, of course, do far more than produce the City of Johannesburg. Yudelman (1984, 9) wrote that, 'The major influence behind the telescoped development of modern South Africa – the leap from a fledgling quasi-state to a surprisingly advanced modern industrial state within the space of eighty years – a process that took centuries in Europe – was the South African gold mining industry'. Innes (1984, 69) referred to the development of the gold mining industry as:

> formative, not only in terms of establishing the capitalist relations of production which were to be the basis of subsequent growth in the industry itself, but also in conditioning the form of evolution of wider social relations in the country, including such phenomena as the migrant labour system, the character and form of the state and the system of labour relations.

The analysis here is divided into two sections: 1886–c.1948 and c.1948–2012. There is a clear rationale for the starting date as this was when the main gold-bearing reef of the Witwatersrand was discovered. The rationale for the divide at 1948 is twofold: 1948 is a political watershed as it was the year that the National Party took power in South Africa and introduced its policy of apartheid, but it also roughly marks the commencement of a period in which manufacturing eclipsed mining as the core of the national and local (Johannesburg) economy.

The article brings together a significant existing literature on the mine labour (Wilson 1972, 2001; Crush 1986; Crush, Jeeves, and Yudelman 1991; Yudelman 1984) with work on: the political economy of mining (e.g. Innes 1984) and of Johannesburg (Beall, Crankshaw, and Parnell 2002); and the changing spatial configuration of Johannesburg (Beavon 2004; Tomlinson et al. 2003). It updates this with reference to recent data and analysis on the mining industry and the changing economy of Johannesburg.

1886–1948: The rise of gold and the rise of Johannesburg

The Witwatersrand

When the Witwatersrand gold fields were first opened up in 1886, there was a ready market for gold as the major European economies were tied to a gold standard, and the liquidity of their currencies depended on a ready supply of the metal. The mining of the gold was, however, extremely costly. For although the gold reefs on the Witwatersrand break the surface in small outcrops, they dip steeply into the earth,

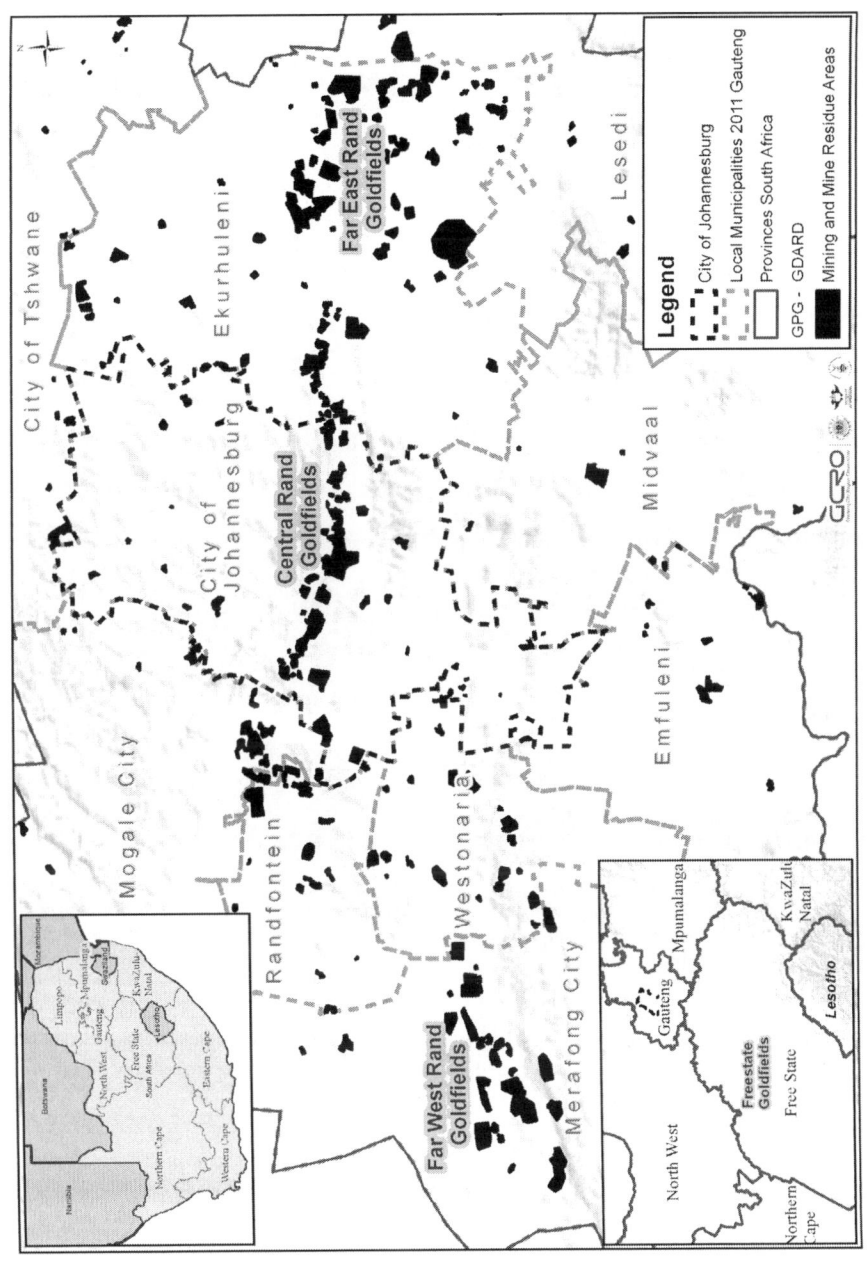

Figure 1. South Africa's gold fields.
Source: Gauteng City Region Observatory (GCRO).

requiring deep-level mining, with expensive technologies, for extraction (Beavon 2004; Innes 1984). It was because of this that a small-scale artisanal mining economy never developed, and that the new gold field was soon dominated by a handful of 'Randlords' who had made their fortunes on the Kimberley diamond fields, and who were backed by international investment capital. These Randlords founded the six dominant mining houses[3] in order to secure the conditions for the continued expansion of mining. A latecomer was the Anglo American Corporation of South Africa, founded in 1917 with capital from the New York bank J.P Morgan. Although there was never to be the same level of concentrated ownership in gold mining as there was with the diamond industry – which was controlled in a near monopoly by the De Beers Corporation – the mining houses established a form of oligarchic control. Ownership became increasingly concentrated as the Anglo American Corporation came to dominate the industry.

Within a decade of the discovery of the gold-bearing reef, the Witwatersrand was the largest gold-producing region in the world. This placed the region at the centre of a major political drama as Great Britain manoeuvred to wrest territorial control of the gold field from the Boer's *Zuid Afrikaansche Republiek* (ZAR) in the Transvaal. The struggle turned violent with full-scale warfare erupting between Great Britain and the two Boer republics in 1899. British forces occupied Johannesburg in May 1900, achieving its objective of economic *and* political dominance of the gold fields. However, in 1906 Great Britain granted autonomy to the Transvaal, which became part of the Union of South Africa in 1910, and the gold fields were thereafter increasingly shaped by South African, rather than British, capital and interests.

Formation of a mining labour force and the South African nation state

There was no guarantee that the gold mining industry on the Witwatersrand would survive. There was a delicate balance between cost and revenue, as the price of gold was fixed, and gold had to be mined at ever deepening levels. White workers arrived mainly from Great Britain and Australia and provided high-level skills, but their labour was expensive. A conventional colonial colour bar protected the interests of these workers by keeping black Africans out of higher-level jobs, but when profits were squeezed, mine owners tried to modify the colour bar and replace white workers with cheaper black African labour (Yudelman 1984).

There was a bitter struggle between white workers and mining bosses when the government agreed to assist the Chamber of Mines in addressing a severe labour shortage after the war by importing around 60,000 Chinese on indentured labour contracts. The Chinese were all repatriated by 1910, forcing the Chamber of Mines to search elsewhere for cheap labour.

From 1896, the Chamber of Mines coordinated the recruitment of black African labour through the Witwatersrand Native Labour Association (WNLA) but it failed to secure an adequate supply of labour from South Africa's 'native reserves' (later the so-called *homelands*). Two-thirds of black African workers in the period from 1910 to 1928 came from the Portuguese East Coast (now Mozambique), but a demand from the Portuguese government that this labour supply be capped compelled the WNLA to recruit further and further afield, in colonial-controlled territories such as Nyasaland (Malawi), Bechuanaland (Botswana), South West Africa (Namibia),

Basutoland (Lesotho), Southern Rhodesia (Zimbabwe) and Northern Rhodesia (Zambia) (Innes 1984; Crush, Jeeves, and Yudelman 1991; Crush and James 1991).

Crush, Jeeves, and Yudelman (1991, 1) wrote that 'there is little doubt that if large numbers of low-wage, unskilled migrant miners had not been recruited from throughout the subcontinent, there would never have been a deep-level gold mining industry in South Africa'. The WNLA's labour recruitment tied much of southern Africa to Johannesburg's burgeoning economy, with the authors describing this as 'South Africa's labour empire' (Crush, Jeeves, and Yudelman 1991, 1). This system of labour recruitment was linked to a pattern of oscillating migration in which male workers were housed in single-sex compounds for limited periods, and returned to families in rural areas when their contracts expired. The state was complicit in supporting this system.

With the extensive use of cheap non-South African labour, mining companies were able to keep the lid on costs for a prolonged period (Wilson 2001). In 1921, however, there was a crisis as the price of gold dropped from 111 shillings per fine ounce to 97 shillings. The Chamber of Mines responded by modifying the colour bar and replacing expensive white labour with black African workers. This provoked a bloody rebellion – in which white workers notoriously marched under the banner 'Workers of the World Unite for a White South Africa' – that was ruthlessly suppressed. Two hundred and fifty people died and the white workers' struggle ended. The so-called Pact Government of the early 1920s pacified white workers by legislating the colour bar but otherwise did not act against mining interests. Throughout this formative period, the state played a critical role in supporting the reproduction of low labour costs to ensure the survival of gold mining, which was so central to the development of the national economy (Innes 1984).

Johannesburg's economic foundations

The spectacular growth of Johannesburg in its early years has been recalled in numerous, mainly romantic, accounts of a mining camp burgeoning into a modern metropolis. A few accounts (notably Chilvers 1948 and JSE 1948) provide a sense of how fragile and contingent this development really was. The initial dependence on gold subjected Johannesburg to the immense volatility of financial speculation.

In 1888/89, there was a great speculative boom in which the average value of mining companies increased five-fold, collapsing towards the end of 1889 when mining companies struck pyritic ore at 100 or so metres below the surface, and had no technical means to extract the gold. Johannesburg seemed doomed to a future as a ghost town. The town was saved in 1891 when a method to extract gold using cyanide was developed. This was followed by a surge in gold shares and another wave of property development, until 1895 when markets crashed because of tensions between the British and the Boers. Mines closed during the war but reopened with great optimism in 1902, only to slump again because of a severe shortage of labour (JSE 1948).

The turbulent roller coaster continued, with peaks of optimism and troughs of despair. In 1930, the future of gold mining and of Johannesburg appeared bleak, with South Africa's Chief Mining Engineer predicting the near collapse of gold mining by 1950 (Shorten 1970). In December 1932, however, South Africa reluctantly followed Great Britain and other major economies in abandoning the gold standard, with extraordinary results. Suddenly freed, the price of gold doubled overnight, and continued to rise over the next five years. There was an 'orgy of speculation in gold shares' (JSE 1948, 89), with massive private profit-taking. The boom incentivised prospecting and exploration, which led directly to the next major wave of mining development in the late 1940s (Chilvers 1948; JSE 1948).

After a peak in gold production in 1941, materials and labour were diverted to the war effort and African labour became increasingly militant (Chilvers 1948). The next speculative boom was in 1945, and again in 1948, when high yields from the newly discovered Free State gold field were confirmed (Chilvers 1948, 262).

In all of this, Johannesburg's role in the physical production of gold was declining. Central Rand's production peaked around 1911, when the gold field accounted for 80% of South Africa's output. During World War I, highly profitable mines were opened on the Far East Rand by the newly established Anglo American Corporation, and from 1923 this new gold field eclipsed Central Rand. During the 1930s, the Far West Rand was opened up, and when the Free State gold fields developed, the relative position of Central Rand declined further. Between 1938 and 1949, Central Rand accounted for 34% of national production (Scott 1951; Viljoen 2009).

Johannesburg nonetheless continued to grow in power, economic size and population because the mining companies, and also the stock exchange, were headquartered in central Johannesburg, channelling profits towards the city. This was reinforced by the rise of the Anglo American Corporation of South Africa, which, unlike the London-based Gold Fields of South Africa, was a South African company with headquarters in the centre of Johannesburg. In the 1940s, Anglo American became a multinational, expanding its interest northwards in Africa, with 51% control of the Zambian copper mines. It also gained ownership of the De Beers Corporation in the 1920s, and so held a monopoly over the diamond industry.

Johannesburg experienced a process of diversification beginning in the early twentieth century. The development of the mines created an immediate demand for industrial production such as iron and steel, explosives, construction materials and chemicals. The first major industrial development happened in 1890 when President Kruger granted a concession for the development of an explosives industry to the Nobel Trust. In 1913, a mining and property development company established Union Steel Corporation as South Africa's first steel producer. During World War I, South Africa faced a disruption to its supplies of imported manufactured goods, spurring the country to manufacture domestically. By the end of the war, South Africa had emergent industries in sectors including electricity, steel, engineering, chemicals, construction materials and clothing, and many of these industries were established in and around Johannesburg, where the mines and the growing population provided a ready market (Innes 1984).

There was a direct and continued involvement in the development of industry by the mining sector. Union Corporation, for example, established the South African Paper and Pulp Industries (SAPPI); Rand Mines spawned the Portland Cement Company; JCI owned South African Breweries (SAB); and Gold Fields operated factories producing clothing, metals, chemicals, construction materials and food. Importantly, the mining industry was also central to the development of an electricity producing and distribution industry. In 1906, the Chamber of Mines supported the establishment of the Victoria Falls and Transvaal Power Company to provide electricity to the mines until the Electricity Supply Commission (ESCOM) was founded in 1922.

Tariff protection introduced by the Pact Government from 1924 significantly aided manufacturing, and there was a further surge in the development of import replacement industry during World War II. At the end of the war, manufacturing was as important to the national economy as mining, with its contribution to GDP having risen from 10% in 1918 to 25% in 1945 (Innes 1984).

A tight link existed between mining and tertiary services. From the beginning mining needed large amounts of capital to finance production, and finance houses quickly emerged alongside the mining houses, with interlocking directorships. Banks and building societies were established in Johannesburg from the 1880s (such as the United Building Society, now part of ABSA Bank) or moved their headquarters to Johannesburg (for example, Nedbank and Standard Bank of SA). The formation of the Johannesburg Stock Exchange (JSE) in November 1887, which supported the equity requirements of the mining industry, was also a critically important development in strengthening Johannesburg's position as the centre of South Africa's emergent industrial and business economy.

Johannesburg's spatial expansion

Johannesburg's changing social structure is not dealt with in detail in this article, but one of the most striking social outcomes of mining that must be noted was the city's changing demographic profile. In 1911, there were only 36 females per 100 males, with 77 per 100 for the white population, and only five per 100 for the black African population. In 1946, towards the end of the first period of study, the overall figure was 75 females per 100 males, with 101 for whites and 56 for black Africans.[4] As sex ratios were narrowing, class differentiation was widening. Van Onselen's *New Babylon, New Nineveh* (1982) provides wonderful detail on the emergence of working class cultures in Johannesburg.

The spatial evolution of early Johannesburg was profoundly shaped by the physical presence of mining and by the hierarchies and intersections of a society that emerged around the mines. Johannesburg quickly emerged as the central node in a string of settlements in a nearly 50 km east–west belt along the gold-bearing reefs of the Witwatersrand. The mining belt was a defining physical feature, with settlement on both sides but mainly to the north as the land to the south was underlain by reefs and reserved mainly for future mining activity.

After the discovery of gold, mining claims were pegged out on privately owned farms along the reef, with mining camps spreading out north and south of the diggings. The formal settlement of Johannesburg was however proclaimed on a triangular piece of leftover state-owned land – *uitvalgrond* – immediately north of the mining belt. The settlement was not considered to be permanent and it was laid out crudely on a tight grid with small blocks (Beavon 2004).

Johannesburg had an improbable location and existed only because of a gold reef with uncertain prospects, but the mining camp soon burst into a bustling town with 'banks, shops, hotels and boarding houses, a stock exchange, and the inevitable saloons and brothels' (Beavon 2004, 6). Within 10 years, Johannesburg was the largest urban centre in Africa south of the equator, and its population of 102,000 exceeded that of Cape Town (Chipkin 1993). The trajectory of growth was not smooth, with the brusque fluctuations of the mining economy, and the effects of war and other disturbances, affecting the rate of population and physical growth. During the slump from 1889 to 1891, one-third of the population left the Witwatersrand, and there was another exodus during the South African War (Shorten 1970). By 1928, however, the government had sufficient confidence in the future of Johannesburg to formally proclaim it a city.

The mining boom of the 1930s brought rapid growth to Johannesburg and dramatic urban transformations. This was the time when the inner city went high-rise. Chilvers (1948, 235) notes that 'Buildings were continually being torn down and replaced by modern skyscrapers. Telephones, power, transport just couldn't keep pace with Johannesburg's development. And the pace was getting faster – and faster'. Alongside the control and repression of black settlement in the city, white space was experiencing a massive building boom as over 10,000 apartments were created on the north-eastern edge of the inner city. The rapid development of manufacturing in the 1940s was linked to the emergence of a string of industrial estates along the mining belt and around the edge of the inner city. The tertiary sector, including commerce and finance, consolidated in the inner city (Tomlinson et al. 2003).

The socially and racially segregated and unequal nature of Johannesburg's development has been the central theme in many accounts of the city's development (Guillaume 2001; Beall, Crankshaw, and Parnell 2002; Murray 2008). Beavon (2004) observes that Johannesburg's geography of segregation was apparent from as early as 1887 and that the patterns that shaped almost all future development were firmly in place by 1904.

By 1904, more than 100,000 black Africans, and large numbers of Chinese, were corralled in regimented single-sex compounds on mining property along the Witwatersrand. Initially the compounds were built of iron and wood, but later there were concrete, barrack-like structures with rooms housing 20 to 50 workers each (Crush and James 1991; Crush, Jeeves, and Yudelman 1991). Not all black Africans, however, lived in these compounds. As black Africans entered employment in other sectors, migrants found accommodation in municipal compounds and also in slums in and around the centre of town and in domestic accommodation in white residential areas. This led to the inter-racial proximity that the city council refused to accept, provoking a long history of attempts to segregate the race groups (Parnell and Mabin 1995). The origin of Soweto, for example, was as early as 1903 when the city council moved black Africans living in a slum in present-day Newtown to a remote settlement 16 km south-west of Johannesburg called Klipspruit, ostensibly in reaction to threats of bubonic plague.

From the time of the Native Urban Areas Act, 1923, which prevented black Africans from purchasing or renting land in white areas, the local authority gradually developed segregated housing estates to which black Africans were moved. The whole of the municipality of Johannesburg was proclaimed white by 1933[5] and by the late 1930s the local authority had used the provision of the Slums Act 1934 to clear mixed-race inner city neighbourhoods, and move black African residents to newly built townships such as Orlando. Much of the new township development was situated at some distance south of the mining belt in the area that became Soweto, thus establishing a fundamental divide in the structure of the city (Tomlinson et al. 2003; Beavon 2004). Indian and coloured (mixed race) communities maintained their foothold near the inner city, in places like Fietas, until the apartheid era, when they too were forced into peripheral townships.

By 1904, the wealthy white elite, including the Randlords, were occupying the high-lying ridges to the north of the mining belt away from the dust and noise of the mines, establishing a pattern of wealthy suburban development in a northwards direction. The white mine-workers who had arrived in Johannesburg from abroad lived in small but solidly constructed bungalows in suburbs strung out along the edges of the mining belt; the Jewish migrants from Eastern Europe and the Russian Empire lived in Yiddish-speaking enclaves in the east of the town; the Afrikaners who had arrived from the farms lived in generally poor neighbourhoods in the west of Johannesburg (Chipkin 1993; Beavon 2004). By the 1930s, the city council developed sub-economic housing estates for 'poor whites', often on land where racially mixed slums had been cleared (Parnell 1988). While extreme forms of segregation are generally associated with the system of apartheid, all of this was happening before 1948. The mining industry was a key driver in the increasing levels of urban segregation, and provided the template for the socio-spatial engineering of the National Party government in later years.

Although the dominant representation of Johannesburg has been of a divided and segmented city, recent literature, informed by post-colonialist theory and cultural studies, directs attention to the cross-over and syncrecity that was also a feature of Johannesburg's development (Nuttall and Mbembe 2008; Nuttall 2009; Bremner 2010). In the pre-apartheid era, there were cultural melting pots where new cultural formations emerged. The British, European and Russian immigrants forged a new English-speaking identity in Johannesburg while a creolised African working class identity arose from the inner city slums. There was also racial mixing that persisted into the 1940s despite the local authority's efforts to separate out the various groups.

1948–2012: The decline of gold and the rise of Johannesburg

Gold and labour force fluctuations

By the end of World War II, South Africa still accounted for 40% of the world's total gold output, with the mines around Johannesburg producing about one-third of the national output. The powerful trends away from mining were, however, already in motion.

In 1944, the Allied Nations signed the Bretton Woods Agreement that reinstated the gold standard and pegged the price of gold at US$35 per troy ounce. Once again South Africa was assured of a steady market for its gold, but with a fixed price and rising costs profit margins came under growing pressure. South Africa's gold production peaked in volume in 1970 when it accounted for 78% of global output, but most mines were economically marginal and the future was uncertain (Viljoen 2009).

In 1971, however, the United States unilaterally left the gold standard and the gold price soared to US$800 by 1980, bringing new prosperity to the South African gold mining industry. In spite of a decline in gold production the revenue earned from gold mining climbed dramatically from R830 million to R10 billion. The 1980s were more difficult. The gold price fluctuated between US$300 and US$500, but

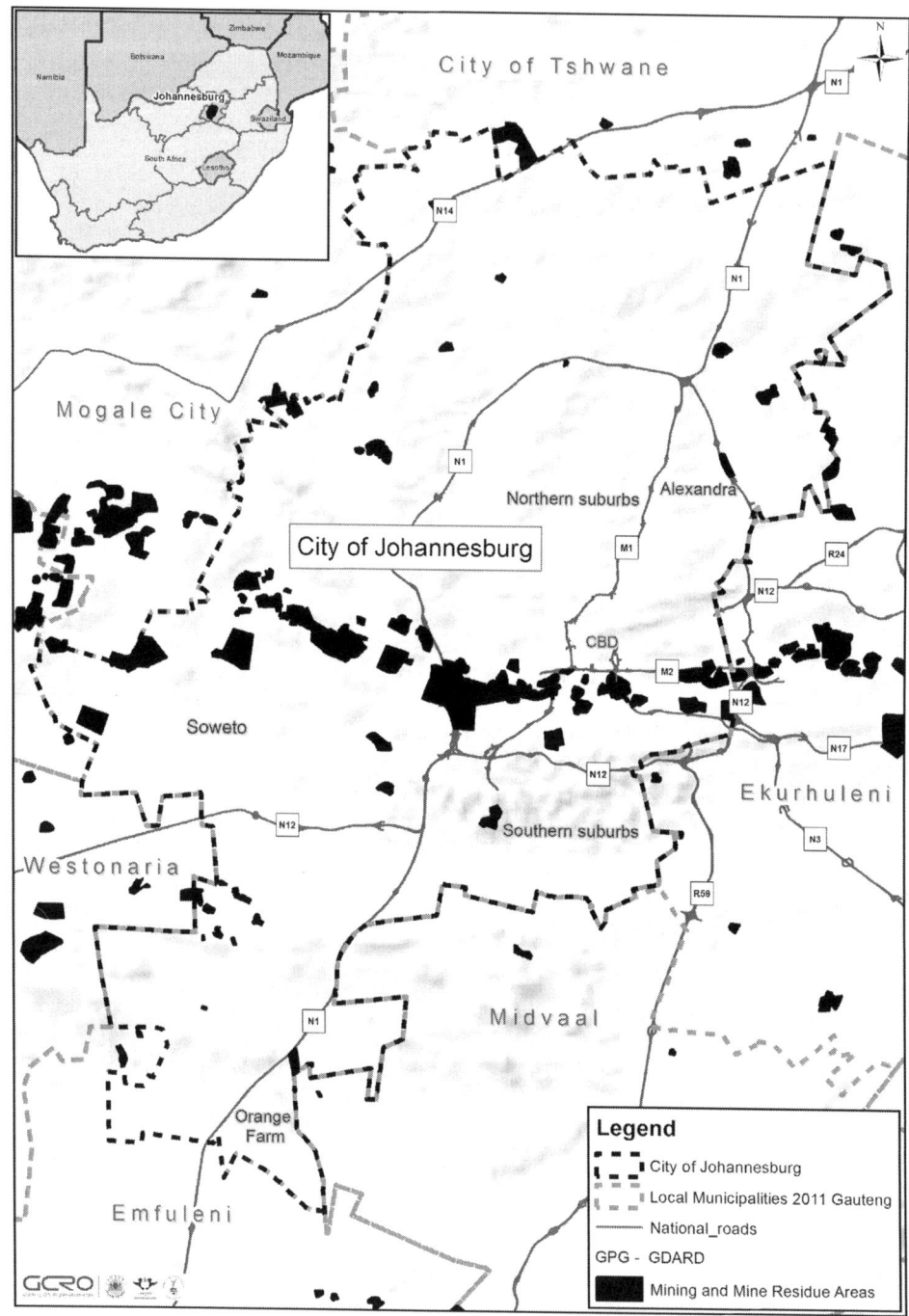

Figure 2. The physical footprint of mining in Johannesburg.
Source: Gauteng City Region Observatory (GCRO).

profits gradually declined, and there was almost no new investment in gold mines (Crush, Jeeves, and Yudelman 1991).

The effects on the labour force were far-reaching. The Chamber of Mines had responded to the cost pressures of the 1960s by expanding the use of foreign labour, with foreign workers accounting for nearly 80% of South Africa's total labour force in 1973 (Innes 1984). In the mid-1970s, however, there was a dramatic change as the colonial government in Mozambique collapsed, abruptly ending labour supply, and as the government of Malawi ordered 120,000 mineworkers to return home (Crush, Jeeves, and Yudelman 1991).

Increasingly, mines had to draw on local labour, and the number of South Africans employed on the mines rose from 87,000 in the mid-1970s to 333,000 in the mid-1980s. Deteriorating conditions in rural South African homelands and the stagnation of the manufacturing economy ensured that there was a strong supply of South African workers. For the first time in the history of gold mining in South Africa there was a surplus of labour (Wilson 2001).

Apartheid was in its final stages, as evidenced by a number of reforms, including: a rapid increase in the real wage of mine workers in the 1970s; recognition for the National Union of Mineworkers (NUM) in 1980; the abolition of the system of influx control into urban areas in 1986; and the lifting of the colour bar in 1988. With mechanisation of the mines there was more demand for skilled labour, and so there was growing talk of the need to 'stabilise' the workforce (Crush, Jeeves, and Yudelman 1991). The system of mining compounds and of oscillating migration did not end, however, with the lifting of influx control. Gold mining companies generally did not follow De Beers Corporation, which had abolished single-sex compounds on diamond mines and provided family accommodation. Instead there was a gradual loss of control over the compounds as families moved in with male workers or found accommodation in burgeoning informal settlements.

Johannesburg and the New South Africa

When South Africa made its transition to democracy in the early 1990s, it was still the world's dominant gold producer, accounting for 44.5% of global output, but difficult days were ahead. In 2010, South Africa produced only 191 tons of gold, 7% of global output, slipping to fifth in the world production rankings (Chamber of Mines 2011).

In the 1990s, mining employment in South Africa fell by 40%, and a further 179,000 jobs were lost between 2001 and 2011. In February 2012, mining production had hit a 50-year low, with the mining sector accounting for less than 5% of GDP. The sector had not only failed to benefit from the commodities boom of the 2000s but was contracting sharply. Gold was worst hit, with a 7% annual contraction in production since 2000, and an 11% decline in 2011 (*Business Times,* 9 June 2012). This deeply constrained the ability of the state to extract rent from the mining sector.

This slump in gold mining has had a severe impact on the economies of gold-producing areas, including towns in the Far West Rand and Free State, but has had no apparent effect on Johannesburg, where other sectors had long since replaced mining. In one sense, the situation in Johannesburg was worse than nationally as the Central Rand Goldfield declined more quickly than elsewhere. By the 1960s, the average profitability per ton of rock mined was only R1.92 for the Central Rand compared with R5.48 for the Far West Rand and R5.59 for the Free State (Cockhead 1970). All the large mines operating on Central Rand[6] shut down by the late 1970s, bringing production on the gold field to a near halt, although the high gold price did allow for the retreatment of old mine dumps (Viljoen 2009). Beall, Crankshaw, and Parnell (2002) calculated that gold mining's share of total employment for Johannesburg fell from 23% in 1946 to 1% in 1996. Very recently, new technologies have made ultra-deep mining feasible and restored the prospect of mining in the Central Rand, but to date efforts to implement the new technologies have not been successful (Seccombe 2012).

To understand the paradox of Johannesburg's continued economic growth despite the collapse of the Central Goldfield, it is necessary to look at the city's position as the economic hub of South Africa's national economy since 1948, which entails its continued role as the corporate rather than physical centre of mining and its continued urban economic diversification supported by the mining companies. The Johannesburg-based Anglo American Corporation, in particular, wielded immense economic power and political influence in South Africa, and was extending its reach internationally (Innes 1984).

In 1999, however, Johannesburg's pre-eminent position in the mining world experienced a setback when Anglo American merged with Minorco and transferred its headquarters and primary stock exchange listing to London. Anglo American did, however, retain a strong presence locally and Johannesburg remains the headquarters of at least seven companies in the Mining Top 100. Johannesburg thus remains a prominent node within a global corporate network of mining firms.

Johannesburg's economy diversified rapidly in the post-war era. In the 1950s and 1960s, manufacturing was the sector that most obviously led growth. Although this put pressure on wage rates in mining, the large mining companies played a dominant role in supporting manufacturing. Anglo American, for example, increased its manufacturing interests in the 1960s by 470% (Innes 1984).

The mines created original demand for industrial products, but a new process was underway in which mining companies expanded into multi-sector conglomerates, with industrial interests not necessarily linked to mining activity. From the 1970s, however, South Africa's manufacturing sector stagnated, and attempts at resuscitation by promoting export-led industrialisation largely failed.[7]

Specialised services, especially finance, took over from manufacturing as the lead sector in the South African economy. Here, too, mining companies played a leading role, becoming a central part of an expanding chain of financial power. Anglo American had significant equity stakes in major banks (Nedbank, Barclays and Standard) and in other financial institutions (such as Eagle Life Assurance) (Innes 1984).

Johannesburg's economic and spatial restructuring

Johannesburg remained at the centre of these economic transformations. Its own economy went through a profound process of change. Mining continued its seemingly inexorable decline. Manufacturing employment grew until around 1980 and then declined sharply thereafter, with its share of Johannesburg's total employment dropping from 24% to 13% in 1996 (Beall, Crankshaw, and Parnell

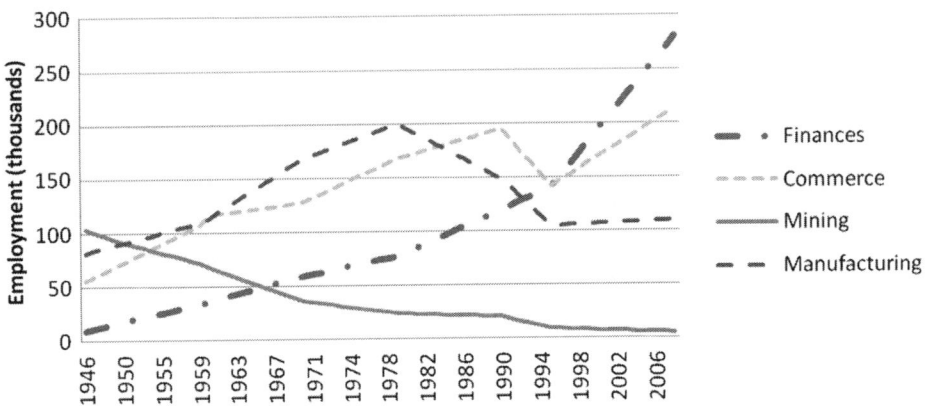

Figure 3. Change in employment by sector 1946–2009, Johannesburg.
Source: Beall, Crankshaw, and Parnell (2002), Quantec.
Data sources: Beall et al. (2002); Quantec

2002; Rogerson and Rogerson 1995). Community, personal and social services grew steadily until around the 1990s and then experienced modest employment decline. It was finance, insurance and real estate that continued growing and eventually outperformed all other sectors.

Initially, the rise of manufacturing more than compensated for the decline of mining and Johannesburg's economy performed well. In the 1980s, however, the decline of both manufacturing and mining took its toll, and Beall, Crankshaw, and Parnell (2002, 33) report 'a negative average annual rate of growth per capital gross geographic product per capita GGP of minus 0.6 percent'.[8]

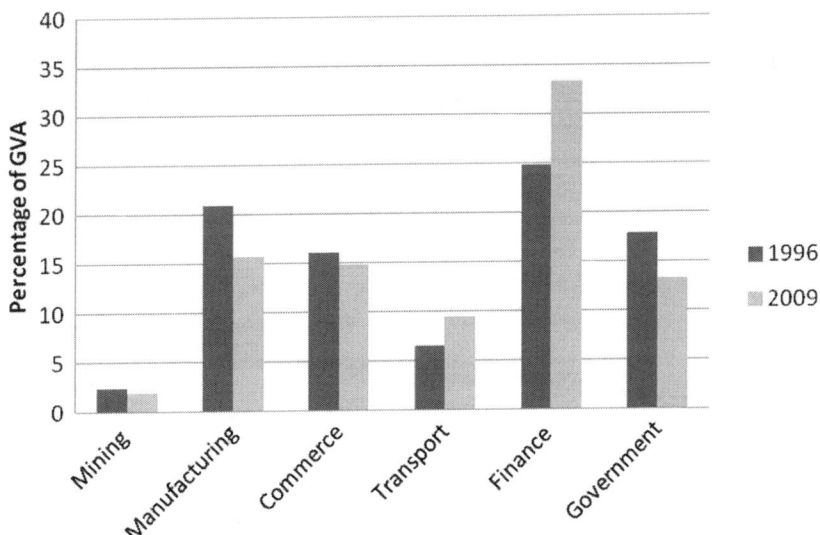

Figure 4. Change in contribution to GVA by sector 1996–2009, Johannesburg.
Source: Quantec.

In the period 1996 to 2009, however, Johannesburg's economy performed well. The local economy grew at an annual average of 4.5% measured in gross value added (GVA). This was higher than the national average of 3.3% and higher than that of any other metropolitan area in South Africa.[9] Johannesburg was consolidating its economic dominance of South Africa.

As Figure 4 shows, mining's direct contribution to GVA in the democratic era is almost negligible, and manufacturing has continued its relative decline, although there has been modest absolute growth. Finance has had the most impact on growth, with its contribution to GVA rising from 24.8% in 1996 to 33.4% in 2009. Johannesburg has emerged as a global service centre in banking, finances and related services (Hamilton 2006). Sassen (2009) ranks Johannesburg as the 22nd city in the world in the 'financial dimension', with a ranking of 8 for 'derivatives contracts' and 10 for 'commodities contracts'.

Present-day Johannesburg, with the economic structure and growth performance of a successful post-primary and post-industrial city, is far removed from its mining origins. However, Johannesburg remains deeply connected to its mining past, with its new flagship sector, finances, having evolved from the financial needs and power of mining.

Spatial transformation of the apartheid city

Although the broad pattern of Johannesburg's spatial pattern may have been established in the early decades, political and economic transformations have had far-reaching effects on spatial configuration. When the National Party took power in 1948 Johannesburg was already a highly segregated city as the city council had cleared most of the racial mixed neighbourhoods by the end of the 1930s and the workforce in the mines was extremely segmented. Apartheid was ruthless and extreme in separating out the remnants of integration, although it ultimately failed in its objective of absolute segregation.

The late 1940s to the early 1970s were the heyday of apartheid and a period of rapid economic growth underpinned by the expansion of manufacturing. This was accompanied by the suburban expansion of white residential areas in the north and south of the city. White suburbia initially developed in areas to the south of the city, ultimately becoming white working class neighbourhoods. The locational impetus for these settlements came less from mining and more from the presence of iron and steel works. It was in this period that the apartheid government developed large segregated townships with no industrial or commercial base for black Africans – most notably the agglomeration of townships that became known as Soweto – and eliminated racially mixed cultural melting pots such as Sophiatown and Western Native Township.[10]

The movement of Africans to the city continued apace and had reached 455,000 by 1948. The workforce for the mines remained in the compounds but there was an involvement of the mining industry in township development. In the 1950s, the founder and head of Anglo American, Sir Ernest Oppenheimer, provided a low-interest loan to Johannesburg City Council to construct 50,000 housing units in Soweto for families who were then living in shanty towns and emergency camps (Beavon 2004). This initiative, which accelerated the development of Soweto, may have represented an early attempt by the mining sector to create a more stable

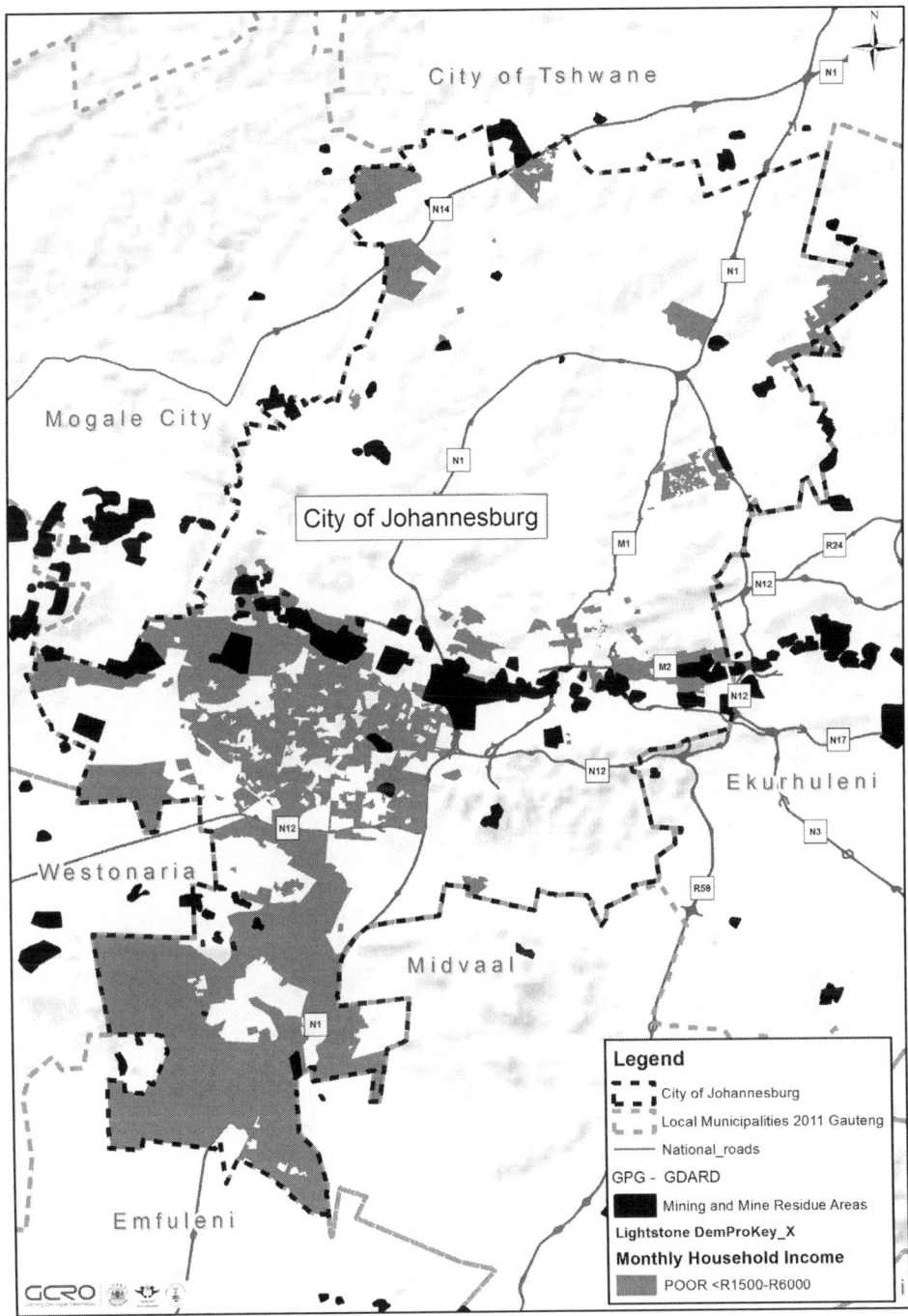

Figure 5. The mining belt in relation to areas of poverty in Johannesburg.
Source: Gauteng City Region Observatory (GCRO).

environment for labour, but it was not followed up in subsequent years. Mining compounds remained almost exclusively the residence of black African mineworkers until the arrival of democracy in the 1990s.

While the state was regulating and structuring the life of black residents, the rapid economic growth – which primarily benefited the white middle class – coupled with the expansion of freeways, supported suburban sprawl to the north of the city. There was a rapid expansion of industrial land around the inner city and along the mining belt as well as near townships with labour supply such as Alexandra. The tertiary sector was also growing, although not yet as dominant as manufacturing, concentrated in the inner city until the 1970s.

From about the time of the Soweto uprising in 1976, Johannesburg entered a new phase in its physical development. With the government having halted its large-scale township development and the system of influx control collapsing, informal settlements and backyard shacks reappeared in the city. After the abolition of influx control in 1986 the government tried to direct new settlement by black Africans to newly created townships such as Orange Farm and Diepsloot on the urban edge as African urbanisation accelerated.

The shift in economic structure from a manufacturing base to the services and financial sectors led to a more diffused spatial form that related to commercial and office decentralisation beginning in the late 1960s. Johannesburg became a sprawling multi-modal urban agglomeration, with private investment happening mainly in the prosperous north. Sandton emerged as the centre of the growing financial services sector, eventually eclipsing the inner city as the core of private sector enterprise in the city.

With the closure of the mines, new land was released for development, but the toxicity of the land and the large number of slimes dams and mine tailings was a major constraint. The largest landowner on the segment of the mining belt in Johannesburg, Rand Mines, established a property development company, which gradually reclaimed land for residential, industrial and business purposes.

Johannesburg in democratic South Africa

The democratic era, from 1994, brought a series of complex transformations to Johannesburg, which represented both change and continuity. As formal business left for the expanding nodes in the northern parts of the city, the private sector invested heavily on the northern frontier, creating new 'edge cities'. Meanwhile, there were dramatic land use and demographic shifts in the inner city. Informal traders and new immigrant communities from across Africa established a large presence. Against the background of pent-up housing demand in black townships and the rise in both internal and sub-Saharan urbanisation to the city, there was an influx of people occupying abandoned inner city buildings and infrastructure, resulting in over-crowded and distressed living spaces. The first decade of the twenty-first century saw many of these buildings converted to low- to middle-income housing opportunities. The inner city encapsulated the competing forces and complexities of gentrification, decline, housing opportunity and exclusion. At the same time, an emerging black middle class left the townships and moved into previously white suburbs, while state-subsidised housing, largely on the edge of the city, created new ghettoes of poverty.

The physical legacy of mining was reasserted during the post-apartheid period. The continued presence of the compounds was a bitter reminder of the past, with government committed to demolishing them or refurbishing them as family accommodation. Mining companies also agreed to this in the Mining Charter 2003. Implementation of the programme was slow and halting as it competed with the multiple demands for state housing subsidies. While some progress has been made with the 'hostel eradication programme', a number of large hostels remain along the old mining belt.

The rehabilitation and redevelopment of the mining belt accelerated demand for industrial, commercial and residential land, renewing growth of the city since the 1990s. The city administration has proposed an east–west corridor of development to knit together the wealthier north and poorer south, but the development has been driven mainly by the strategies of the mining companies. There is a lingering dispute over the right of a municipality to impose planning controls on mining land, but a recent judgment of the High Court in the Western Cape upheld the right of municipalities to regulate land development on all land within their jurisdiction (Kidd 2011). This, together with a plan to consolidate all mine tailings along the mining belt at a single site, may lead to more coherent land development away from the extremely disjointed patterns at present.

A deep concern currently is the danger of acid mine drainage arising from previous gold mining activity. As water percolates into old mines, slimes dams and dumps it is exposed to pyrites and other sulphides and becomes acidic. As water pumps no longer operate, this acidic drainage gradually rises and eventually spills onto the surface, threatening vegetation, water courses, human health and even the foundations of buildings. There have been belated efforts by government and mining companies to address this threat but the levels of acid mine drainage are still rising and have reached the surface in places (Naicker, Cukrwosja, and McCarthy 2003; Kidd 2011).

Conclusion

In common with a handful of other cities which have survived the boom–bust cycle of a mineral economy, Johannesburg owes its origins to mining. What is also important, however, is the *contingency* of Johannesburg's development. The mining industry in Johannesburg, shaped by considerations ranging from geotechnical to geopolitical, took a political form that had immense consequences for the future development of the city, but also for the shaping of social, political and economic formations and relations across southern Africa and beyond.

The minerals revolution that followed the gold discoveries fuelled far-reaching processes of urbanisation; shaped the nature of labour relations in southern Africa for more than a century; tied the development of communities from the rural parts of South Africa and neighbouring countries to the requirements of mining on the Witwatersrand; and provided the template for socio-spatial engineering of the apartheid government. Johannesburg was originally built to serve the mining industry and this is reflected in its spatial structure.

Mining has faded but Johannesburg has continued growing. Today Johannesburg can hardly be regarded as a mining city. However, the industries and sectors that now dominate Johannesburg's economy are themselves a product of a mining history.

Johannesburg's current status as an emergent global service centre in finance owes much to the historical links between mining and banking.

The extent to which the ills and the virtues of the large and complex urban agglomeration, of which Johannesburg is the core, is owed to mining is poorly understood and rarely acknowledged. In recent years, however, the ghost of the mining past has reappeared with the environmental threats posed by acid mine drainage, and also with some renewed prospects for mining within the Central Rand Goldfield. The future is uncertain but gold mining is unlikely to emerge again as a major economic activity. Johannesburg's economic prospects rest mainly on its ability to consolidate its role as a financial and service centre at a time of global and national economic vulnerability.

Acknowledgements

The authors acknowledge the support of the South African Research Chairs Initiative (SARChI) of South Africa's National Research Foundation (NRF) and the valuable assistance of Jennifer Paul in producing the maps and of Yan Yang in preparing the tables.

Notes

1. Beavon (1997, 153) first wrote that 'gold . . . ignited the development of Johannesburg'.
2. Figures provided in Viljoen (2009).
3. Consolidated Gold Fields of South Africa, Rand Mines, General Mining, Union Corporation, and Johannesburg Consolidated Investments (JCI) collaborated through the powerful Witwatersrand Chamber of Mines.
4. See the National Population Census reports of 1911 and 1946 available in the national government and legal deposit libraries in South Africa.
5. In response to resistance from black freehold landowners and because the municipality could not provide alternative accommodation for all the Africans that it would need to house, the adjoining areas of Sophiatown, Newclare and Martindale were exempted from the declaration (Beavon 2004).
6. From west to east the mines were: Durban Roodepoort Deep, Rand Leases Mines, Consolidated Main Reef, Crown Mines, Robinson Deep, City Deep, Simmer and Jack.
7. The cause of this decline in performance has been extensively debated, with observers pointing to racial policies which prevented the growth of a black middle class and so restricted the size of the domestic markets, declining productivity and protectionism which prevented a competitive export market from developing (Bell 1995; Kaplan 2010).
8. Comparability of statistics is a problem. Figures for GGP were only provided between 1968 and 1991. Since 1996, private firms (Quantec and Global Insight) provide modelled estimated of economic output for municipalities for South Africa but these boundaries do not coincide with the previous statistical boundaries.
9. Based on figures provided by Quantec.
10. Segregated townships were also built for the Indian (Lenasia) and coloured (Eldorado Park and Ennerdale) groups.

Notes on the contributors

Philip Harrison is a Professor holding the South African Research Chair in Development Planning and Modelling hosted by the School of Architecture and Planning at the University of the Witwatersrand, and is a member of South Africa's National Planning Commission. He has worked in various academic positions, including at the University of KwaZulu-Natal, and has also held positions in the private and government sectors, including as the head of planning and urban management in the metropolitan city of Johannesburg. He has co-authored/co-edited two books relating to urban development, and is currently heading a research programme that explores processes of urban spatial transformation in large cities. He can be contacted at: philip.harrison@wits.ac.za

Tanya Zack is a Town and Regional Planner with 20 years' experience in municipal, private and non-governmental environments in housing, planning and development issues in Johannesburg. She holds a PhD from the University of the Witwatersrand for her application of critical pragmatism in the evaluation of planning practice. Her recent work in Johannesburg includes the development of policy responses to derelict and distressed buildings, the review of a public–private Inner City Charter, a short-term role in the executive of the Johannesburg Development Agency, responsible for area upgrades in the city, as well as the examination of migrant space in the heart of the inner city. She is a Visiting Researcher at the University of the Witwatersrand, Johannesburg. She can be contacted at: tanyazack@icon.co.za

References

Beall, J., O. Crankshaw, and S. Parnell. 2002. *Uniting a divided city: Governance and social exclusion in Johannesburg*. London: Earthscan Publications.

Beavon, K. 1997. Johannesburg: A city and urban area in transformation. In *The urban challenge in Africa: Growth and management of its large cities*, ed. C. Rakodi, 150–91. Tokyo: United Nations University.

Beavon, K. 2004. *Johannesburg: The making and shaping of the city*. Pretoria: University of South Africa Press.

Bell, T. 1995. Improving manufacturing performance in South Africa: A contrary view. *Transformation* 28: 1–34.

Bremner, L. 2010. *Writing the city into being: Essays on Johannesburg, 1998–2008*. Johannesburg: Fourthwall Books.

Chamber of Mines of South Africa. 2011. *Annual Report, 2010/11*. Johannesburg.

Chilvers, H. 1948. *Out of the crucible*. Johannesburg: Juta and Co.

Chipkin, C.M. 1993. *Johannesburg style: Architecture and society 1880s–1960s*. Cape Town: David Philip.

Cockhead, P.J. 1970. The East Rand: A geographical analysis of the transition of the economic base of the region from gold mining to manufacturing, and its effects upon future economic development. MA thesis, University of the Witwatersrand.

Crush, J. 1986. The extrusion of foreign labour from the South African gold mining industry. *Geoforum* 17, no. 2: 161–72.

Crush, J., and W. James. 1991. Depopulating the compounds: Migrant labour and mine housing in South Africa. *World Development* 9, no. 4: 301–16.

Crush, J., A. Jeeves, and D. Yudelman. 1991. *South Africa's labor empire: A history of black migrancy to the mines*. Cape Town: David Philip.

Davis, D. 1998. The minerals sector, sectoral analysis, and economic development. *Resources Policy* 24, no. 4: 217–28.

Guillaume, P. 2001. *Johannesburg – Geographies de l'exclusion*. Paris and Johannesburg: Karthala/IFAS.

Hamilton, M. 2006. Global cities of the South: Mexico and Johannesburg in an era of globalization. MA thesis, University of Stellenbosch.

Innes, D. 1977. The mining industry in the context of South Africa's economic development, 1910–1940. Seminar Paper, Institute of Commonwealth Studies, London.

Innes, D. 1984. *Anglo American and the rise of modern South Africa*. Johannesburg: Ravan Press.

JSE (Johannesburg Stock Exchange). 1948. *The story of the Johannesburg stock exchange: 1887–1947*. Johannesburg: Committee of the Johannesburg Stock Exchange.

Kaplan, D. 2010. Manufacturing in South Africa over the last decade: A review of industrial performance and policy. *Development Southern Africa* 21, no. 4: 623–44.

Kidd, M. 2011. Developments in South African environmental law during 2010: Focus on two decisions relating to mining, IUCN Academy of Environmental Law, e Journal Issue 2011, 1. http://www.politicsweb.co.za/politicsweb/view/politicsweb/en/page71656?oid=196410&sn=Detail.

Murray, M. 2008. *Taming the disorderly city: The spatial landscape of Johannesburg after apartheid*. Cape Town: UCT Press.

Naicker, K., Cukrowoska, C., and T.S. McCarthy. 2003. Acid mine drainage from gold mining activity in Johannesburg, South Africa and environs. *Environmental Pollution* 122: 29–40.

Nuttall, S. 2009. *Entanglement: Literary and cultural reflections on post-apartheid*. Johannesburg: Wits University Press.

Nuttall, S., and A. Mbembe, eds. 2008. *Johannesburg: The elusive metropolis*. Johannesburg: Wits University Press.

Parnell, S. 1988. Public housing as a device for white residential segregation in Johannesburg, 1934–1953. *Urban Geography* 9: 584–602.

Parnell, S., and A. Mabin. 1995. Rethinking urban South Africa. *Journal of Southern Africa Studies* 21, no. 1: 39–61.

Rogerson, C., and J. Rogerson. 1995. The decline of manufacturing in inner city Johannesburg, 1980–1994. *Urban Forum* 6, no. 1: 17–42.

Sassen, S. 2009. Cities in today's global age. *SAIS Review* 1: 3–32.

Scott, P. 1951. The Witwatersrand goldfield. *Geographical Review* 41, no. 4: 561–89.

Seccombe, A. 2012. Central Rand goes back to basics. *Business Day*, 16 April.

Shorten, J. 1970. *The Johannesburg saga*. Johannesburg: John Shorten Proprietry Limited.

Tomlinson, R., A. Beauregard, L. Bremner, and X. Mangcu, eds. 2003. *Emerging Johannesburg: Perspectives on the post apartheid city*. New York and London: Routledge.

Van Onselen, C. 1982. *New Babylon, new Nineveh: Everyday life on the Witwatersrand 1886–1914*. Johannesburg and Cape Town: Jonathan Ball Publishers.

Viljoen, M. 2009. The life, death and revival of the Central Rand Goldfield. *World Gold Conference* 2009: 131–8.

Wilson, F. 1972. *Labour in the South African gold mines, 1911–1969*. Cambridge: Cambridge University Press.

Wilson, F. 2001. Minerals and migrants: How the mining industry has shaped South Africa. *Daedalus*, 130, no. 1: 99–121.

Yudelman, D. 1984. *The emergence of modern South Africa: State, capital and the emergence of organized labour on the South African goldfields, 1902–1939*. Cape Town: David Philip Publisher.

Mining, welfare and urbanisation: the wavering urban character of Zambia's Copperbelt

Patience Mususa

This article focuses on the character of life and social welfare services in the mining towns of what was once the most urbanised country in central Africa. The services provided by mining companies varied over the years: from minimal at the time of the industry's establishment in the 1920s; to a period of largesse between the 1950s and the late 1970s; and then a slow decline following the slide in world copper prices. The withdrawal of the mines from welfare provision from the mid-1990s to the present has radically altered not only people's well-being, but also the character of the urban areas, leading to the observation that towns have become like 'villages'.

Zambia's changing fortunes in copper mining are mirrored in its urban growth and welfare trajectory. The fluctuating world market price for copper has had a direct bearing on the urban population's levels of material welfare and deprivation. The analysis is divided into two main parts: a historical review of mining's impact on urbanisation followed by an examination of the current urban and welfare circumstances of the population following the privatisation of Zambia Consolidated Copper Mines (ZCCM).

Urbanisation on the Copperbelt since the mines came into production in the 1930s was synonymous with rural to urban migration, which was spurred on by the idea of a better 'modern' life in the towns. Several studies from the 1990s suggest that the decline of the industry affected urban processes. These argue that Copperbelt residents were moving from urban to rural areas and settling and seeking subsistence livelihoods in the 'bush' on the outskirts of Copperbelt towns (Ferguson 1999; Hansangule, Feeney, and Palmer 1998; Potts 1995). The 2010 census indicates that Copperbelt residents were moving to the capital city, Lusaka, which would tally with the perceptible shift of Lusaka's lingua franca from Nyanja towards Bemba.

Furthermore, data from the 2000 census shows shifts in regional livelihood diversification patterns with the economically active in mining falling from 3.4% of the country's total labour force in 1990 to 1.3% in 2000 while those active in agricultural activities grew from 50% in 1990 to 72% in 2000 (CSO 2003). On the other hand, since approximately 2004, some rural areas are emerging as small-scale mining outposts in the new mining area of the North Western Province. They are becoming more 'town-like' as former subsistence farmers and semi-foragers turn to small-scale mining joined by 'old' Copperbelt in-migrants and others from as far away as Lusaka who are moving in to benefit from a boom in copper.

While these trends have not changed the cosmopolitan nature of the Copperbelt, which, from its early years attracted a migrant population composed of various African groups and transient Europeans, they have influenced Copperbelt resident's perception of the 'urban'. This has been in the context of the acute urban poverty that followed the mines' privatisation, retrenchments and withdrawal of social welfare provisions. Copperbelt residents complain that their urban areas have lost their sense of 'urban order'.

These changes have laid bare the fallacy of the idea of the urban as modern, which the former neat façade of the Copperbelt towns could elude. Myers (2003, 56), when considering African colonial cities, notes that the planned ideal of the modern city never materialised as these places 'became reframed within the African idioms of urban life, dependent on uneven and unequal development of power by individual householders, or religious institutions, or ideas of space and on customary neighbourly understanding'.

Edward Casey (1997) argues that space should be seen in its 'two-ness', one that colours the temporal experience of being or dwelling in a 'place', and the other that traces paths, from, to and beyond this temporality. This means thinking about the Copperbelt not only as an extractive locale for copper whose activities are affected by the market, but also as a place where the residents' engagement with the reality of losing jobs, and struggling to earn a living amidst the withdrawal of mine welfare, is re-texturing simultaneously the material and social character of the place. What this implies is a blurring of boundaries in activities that connote the modern, and thus urban: for example, the formal copper mining economy, and those implied by a subsistence livelihood, the village.

A term that well embodies this process of the urban becoming like a village, is villagisation, defined by René Devisch (1996, 573) as 'a process of psychic and social endogenisation of modern city life, that allows the migrant to surmount the schizophrenic split between traditional, rural and "pagan" life as against the new urban Christian world'. It is a process that allows for several possibilities of living in the environment, and resists the tendency towards linearity of urban narratives. It enriches what Ferguson (1999, 221) observed to be a situational stylisation of social life on the Copperbelt that drew on idiomatic identification of persons as being *uwa kumushi*,[1] 'from the village', or *uwa kutown*, 'from the town'; to be more like Filip de Boeck's (1998) observations of the Aluund people and their pragmatic relationship to the environment in the context of hunger in the Democratic Republic of Congo. Van Binsbergen's (1998) concept of the virtual village refers to people who avoid identifying solely with the town or the village. This pragmatism also enters Copperbelt residents' conception of themselves and their position within a changing world. Residents practise 'converged' lifestyles with often internally discordant modes of social interactions as described in this article.[2]

Following Zambia's independence in 1964, President Kaunda tried to address the relation between the rural and urban through his philosophy of Humanism, attempting to reframe urban life by transposing 'African' values based on the idealisation of village life, into the towns; and encouraging an African conviviality of giving across extended urban and rural kin, devolving the welfare of the state in urban areas to the rural through the 'family' (Kaunda and Morris 1966). This theme informed Zambian state policies for the first few decades of national independence when copper underlined the country's economic prosperity.

Copper: historical conduit of urbanisation and social welfare

Emergence of the Copperbelt

Artisanal copper mining had long been practised in the region between the Zambezi and Congo basin. Nineteenth and early twentieth century European prospectors' accounts attest to extensive ancient copper working and they speculated that this had flourished during the slave trade and thereafter became dormant with the abolition of slavery (Bradley 1952). This suggests a centuries-long history of copper production, affected by dips and rises in international trade.

Industrial copper mining in Zambia was facilitated by Cecil Rhodes's acquisition of mineral rights through the Lochner concession, which was held by the British South Africa Company (BSAC) and covered most of the western half of Northern Rhodesia/Zambia. The confirmation of copper deposits by Frederick Burnham, a scout of the BSAC, was the catalyst for mining development. The nature of the ore

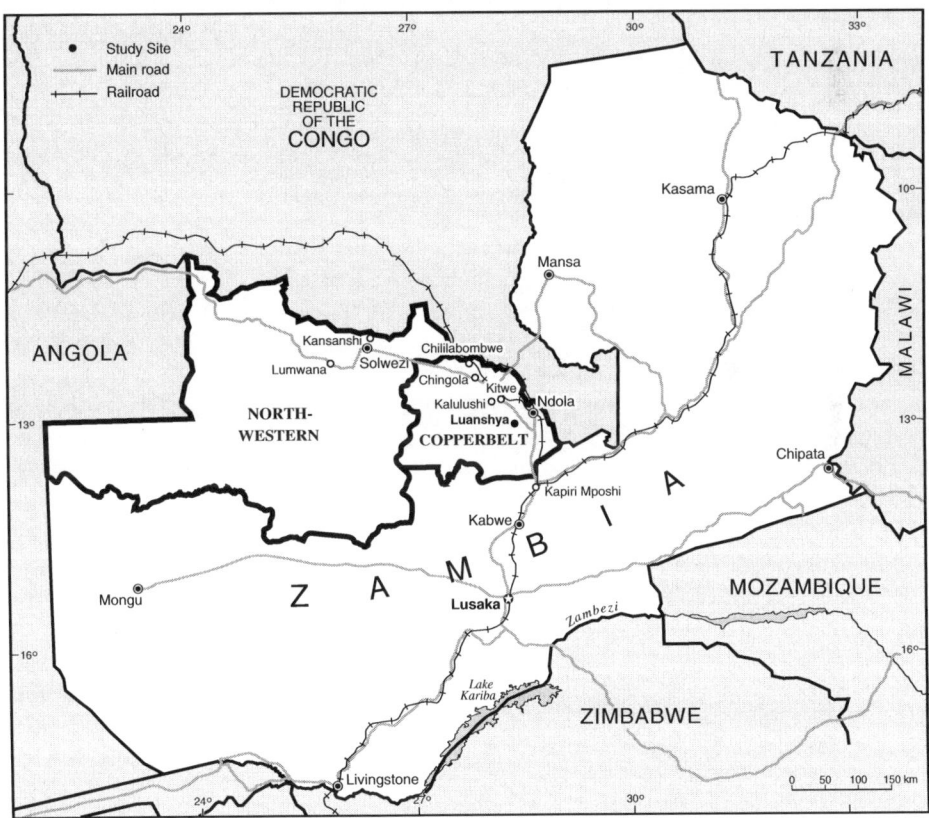

Figure 1. Map of Zambia.
Source: Cartography by M. Shand, University of Glasgow, based on United Nations Population Division map, http://www.un.org/Depts/Cartographic/map/profile/zambia.pdf.

deposits necessitated deep-shaft mining and the lack of a transport infrastructure to take the minerals to ports delayed investment in the area for a decade (Coleman 1971).

The construction of a 'Cape to Cairo' rail line reached Ndola in 1909, creating a corridor along which towns with an urban character were destined to develop. The BSAC's concession was exploited by the two mining houses, Rhodesian Selection Trust (RST) and the Anglo American Corporation (AAC), which were dominant until the early 1970s. Large-scale commercial farming did not emerge in the area due to the acidity of the soils of the copper mining region, as Clifford Darby (1931) noted already in the 1930s. Thus there was a stark contrast between the urban mining strip and the rural subsistence farming zone that was adjacent to it.

Coalescing urban growth and welfare awareness in the Copperbelt, 1909–1939

Many of the early African migrant mine workers stayed in unhealthy camps, which, in the case of Luanshya, were plagued by malaria and black-water fever associated with swampy areas (Schumaker 2008). By the 1930s some of these camps had been improved by the efforts of public-health experts, engineers and vanguard town planners, who were influenced by the garden-city movement and endeavoured to apply these ideas to build small, habitable mining towns. The towns that developed around mining activity included Kabwe (formerly Broken Hill) in the Central Province, where lead was mined, Luanshya, Kitwe, Mufulira, Chingola, Chilila-bombwe, Kalulushi and Ndola, which served as the administrative and commercial centre. As African settlement increased on the Copperbelt, it attracted considerable academic interest (Mitchell 1954; Powdermaker 1962; Epstein 1981, 1992). The parallel setting up of colonial administrative centres resulted in towns with a dual administrative system of the mines and local councils (Mutale 2004).

This early mining period was not characterised by any coherent plans for the welfare of African miners. In fact, the mining companies had a negative attitude towards the permanent settlement of African miners in the towns. They preferred to use male migrant labour, discouraging the presence of African women and children. Despite this attitude, the competition for labour with the more established mines in the region, and the need to feed workers, meant that the mining companies actively supported agricultural activity that was carried out primarily by women cultivators who sold food to the mine workers; and thus in turn implicitly accepted the presence of women (Chauncey 1981). Subsistence plots were provided by some of the mines, first in Broken Hill (Kabwe), then more widely by Roan Antelope mill in Luanshya with 2000 agricultural plots in use in 1935 (Chauncey 1981; Wilson 1968). This stimulus to the surrounding agrarian economy was helpful to the indigenous African population on the Copperbelt and to the unemployed workers who remained in situ during the recession of the early 1930s, which occurred soon after copper production had been initiated in 1929. Their agricultural activity provided a vital subsistence resource and a supplement for miners' families until the ZCCM period.

Centrality of large-scale mining companies as providers of urban welfare, 1940–1960s

Following major strike actions in the 1940s and a commission of inquiry to look into concerns raised by the workers, the mines adopted a welfare orientation. This was at a time when copper was bought at a stable fixed price by the British colonial government to support the Second World War effort. Thereafter, the post-war boom in world demand for copper enabled the resumption of market-based trade. The colonial government's adoption of a welfare orientation was outlined in the first 10-year development plan, which provided for infrastructural developments such as urban housing and other facilities. The implementation of this plan improved living conditions in the towns particularly for both European and African service workers, as housing and recreational facilities were built outside of the mine townships and compounds.

Conditions for the African workforce on the mines also improved as the mining companies, spearheaded by Sir Ronald Prain and RST, built housing for families and provided the opportunity to progress into skilled jobs under a policy of Africanisation. Prain (1956) argued that for the long-term stability of the Copperbelt mining industry and the fostered development of a pool of skilled Africans in the towns, Africanisation was based on the pragmatic policy of employing African workers in skilled jobs to replace European workers who were more costly to hire.

Mining revenue played an important role in funding infrastructural development in the country. During the colonial period, the mining houses paid mineral royalties on copper production to the BSAC and a tax on company profits to the state. After independence, and the expropriation of the BSAC's mining rights, the mining companies paid royalty taxes of 13.5%, export taxes of 40%, when the price of copper went beyond US$300 per long tonne, and corporate tax of 45% to the state (Lungu 2008). This revenue financed the implementation of the five-year transitional development plan that followed the country's independence in 1964. Despite many attempts both before and after independence to distribute this revenue evenly across the country for the development of the sparsely populated and infrastructure-poor rural areas, mining revenue was mainly spent on the urban areas of the Copperbelt Province and the capital city, Lusaka.[3] This use of government revenue for the development of urban areas was remarkable in view of the colonial government's concern that 'over-urbanisation' would result in the rural areas being drained of people, and the urban areas being unable to cope with the influx of large numbers of people (Heisler 1971).

The towns on the Copperbelt, in what many describe as the golden era of the 1950s to the 1970s, had much to offer in terms of the residential lifestyle of those who stayed in the areas administered by the mines. There was decent subsidised housing that had been transformed since the 1940s from African mineworkers' bachelor quarters measuring about 20 square metres with communal ablution facilities to family homes measuring over 50 square metres with indoor plumbing by 1952 (Mutale 2004). For recreation, mine employees had the option of sports clubs, libraries, theatres, cinemas and ballroom dancing facilities, which were also open to their families. Hortense Powdermaker (1962) provides a vivid description of life on the Copperbelt in these times. The mines also had good schools and health facilities, and offered various skills-training programmes, including adult literacy. With regard to their incomes, while the wage disparities between Africans and Europeans that characterised the colonial period prevailed even in the post-independence era, the wages were high enough to enable workers to save towards the purchase of a car. This prosperity contributed to the self-confidence of the country (Fraser 2010).

Mine nationalisation and gradual decline of urban welfare, 1970–1991

Various reasons have been given for the decline of urban welfare in Zambia, and the Copperbelt in particular as world copper prices plunged in the mid-1970s. Firstly, the Matero Economic Reforms of August 1969 allowed the Zambian government to take over the mining companies by buying a majority stake in them and tied politics to economics in a way that reduced the flexibility of the business sector. Partial nationalisation of the RST and AAC mines in 1970, and the cancellation of the mining companies' management contracts in 1973, allowed the state to influence the running of the mines, leading to retention of the labour force even as production was falling due to the drop in world copper prices and reduced investment.

Secondly, the Zambian state failed to diversify its economy, dooming the country to over-reliance on a single commodity, copper. Thirdly, the price of copper was subject to flexible market pricing, making the country and the Copperbelt extremely vulnerable to the rise and fall of the market. Fourthly, following independence, the government failed to create an alternative to the flexible market pricing of copper. A cartel of copper-producing countries, Cipec, was created but it did not succeed in cushioning Zambia from the vagaries of the world market (Fraser 2010). Fifthly, bad timing and poor luck prevented Zambia from benefiting from policies intended to increase state revenue from copper (Fraser 2010). Soon after the partial nationalisation of the mines in 1970 when the state could have benefited from a mineral tax of 51% and 45% corporation tax, a global recession began. This setback forced the Zambian government to reconsider the tax system and offer incentives to investors through a lower tax rate (Lungu 2008).

In spite of these handicaps, the state extended the Zambianisation programme, a continuation of earlier Africanisation policies. The mines expanded in-house job training, built new trades schools for the mines and a technical secondary school, Mpelembe, which in 1983 served a third generation of Copperbelt residents, many of whom went on to study engineering subjects. In an attempt to reduce the wage bill for expatriate workers, on which the mining industry was still dependent, ZCCM built primary schools offering superior quality education in each Copperbelt town so as not to have to pay for schools abroad for expatriate workers' children. If there had been a change of fortunes in the world price for copper, this continued investment in education and building skills could have paid off. However, world copper prices did not improve.

In an attempt to reduce the country's foreign exchange spending, as earnings from copper dwindled, the government scaled up import substitution, encouraging the fabrication of industrial components and consumer products locally. For the wider population a well-remembered example of this was the substitution of locally manufactured soft drinks for Coca-Cola, which unfortunately had a shorter shelf life. Towards the late 1970s, the country began to be besieged by regular shortages of basic commodities like bread, cooking oil and soap as the country's national debt deepened. In the mid-1980s, under pressure from the IMF/World Bank, the state sought to reduce the amount spent on subsidising maize meal for urban residents by introducing a coupon system for the urban poor, and many urban residents had to queue for hours for their supply of maize meal. However, people working for the mines did not have to do so, as they had a separate distribution system, one of several privileges mine employees enjoyed.

Kaunda's failure to see through these reforms was put down to the vulnerability of his government, which had become a one-party system in 1971 under his continued leadership since independence, and was increasingly viewed as politically repressive. However, in 1989, with rising food prices and discontent over the lack of political freedom, popular riots started by the university union on the Copperbelt forced Kaunda to declare early elections in 1991. Frederick Titus Chiluba, leader of the Movement for Multi-Party Democracy, won by a landslide vote, dislodging Kaunda from his long sojourn as president.

Despite the heavy criticism of the mismanagement of the country's economy and mining sector by Kaunda's government between 1970 and 1991, in retrospect many Copperbelt residents remember it as a time when they could at least have a daily meal. A few of the older Copperbelt generation over the age of 50, such as Mr Mubita, a former ZCCM employee resident in Luanshya, attribute the decline of the economy to the departure of the 'whites', with reference to the nationalisation of the mines and, 'too much politicking'. Some of the younger generation, born in the 1970s, who experienced childhood in the lean years of Kaunda's socialist-cum-capitalist experiment, blame the abandonment of the African socialist model for the decline in the economy. Bissell's (2005) cautionary note on nostalgia is valid here in that it is useful to take into account 'multiple strands of remembering'. The years of decline that followed were for the younger generation recalled in the *Lusaka Times* (26 March 2010) article 'Zambia: Those were the days!' as times of simpler pleasures, free from the anxieties of a consumer culture that took root in the privatised era. This generation, never having experienced the golden era, are sometimes critical of privatisation, which is seen as an interruption of the Kaunda government's attempts towards self-determination.

Privatisation policy – 1995 onwards

The Movement for Multi-party Democracy (MMD) government, under the leadership of Chiluba, came to power with a programme of widespread economic reform centred on the privatisation of the mines. After a lengthy debate over whether the mines should be sold and, if so, as a whole, or broken up in parts, they were eventually offered as 'unbundled' units that roughly coincided with mining operations in each mining town. Thus the sale of mining divisions of the ZCCM commenced in 1995 with the sale of the Luanshya mines to the Binani Group. This was not a transparent process, and was dogged with allegations of corruption, which, in the end, besieged Chiluba's administration (Van Donge 2008). In the case of the sale of the Luanshya mines, these dodgy dealings in subsequent years served to obscure ownership of the mines and had a devastating and destabilising effect on the residents of the town (Gewald and Soeters 2010).

During this time, an estimated two-thirds of the ZCCM work force was laid off. Without enough money to pay their retrenchment benefits, the state opted to sell the mine housing stock to sitting tenants, most of them mine employees. A cash benefit package for retrenched miners was, in most cases, paid after a delay of three years. Many Luanshya residents remember this period as one of extreme suffering, with families resorting to eating raw mangoes, foraging for wild fruit in the nearby forests, and turning to the 'bush' to cultivate.

Fraser and Lungu (2007) recount the effect that the sale of the mines had on the mining industry in Zambia. They point to the casualisation of labour and the resultant weakening of the mine-workers union. Larmer's (2007) in-depth historical study demonstrates how the union had hitherto played a central role in lobbying continually for better wages and living conditions on the Copperbelt. Fraser and Lungu (2007) point to the inability of the Zambian state to monitor and regulate the mining industry, a failing that Haglund (2010) notes resulted in numerous abuses that included violations in health and safety and an incoherent investment culture that encouraged patronage. This inability had serious repercussions for the collection of mining revenue.

By way of illustration, *The Post* on 9 February 2011, reported allegations of irregularities in Mopani Copper Mines tax remittances with the falsification of copper revenue to avoid taxes. This was deemed especially underhanded given that mine privatisation had, it came to light, involved secret development agreements that offered very generous benefits to the mines. Corporate tax had been offered at 25% and mineral royalties at 0.6%, zero taxes on customs duties and up to 20-year tax breaks. With rising copper prices from 2004, this tax rate began to seem extremely unfair to the Zambian populace, and under pressure from civil society and the urban-supported opposition party, the Patriotic Front (PF), the government – led by Levy Mwanawasa – was forced to engage in a renegotiation of the development agreements, which resulted in the introduction of a windfall tax in 2008.

Unfortunately for the country, just as this new tax was being put in place, copper prices that in the early half of 2008 had been approximately $8000 a tonne plummeted to $3000 a tonne. The mines, in response to this drop, laid off workers, many of whom were just beginning to feel confident about the economy. In Luanshya, the mines, which had been taken over by JW & Enya in 2004, were placed under care and maintenance, sending the town's residents into a wave of despondency. Around this same time, Levy Mwanawasa died. Following an election held in October 2008, which was strongly contested by the PF, Mwanawasa's vice-president, Rupiah Banda, running on the MMD ticket, was ushered into office, albeit amidst election irregularities (Laterza 2008). The PF had developed a strong following in urban Zambia for articulating social concerns in contrast to the abstract rhetoric of growth of the MMD (Larmer and Fraser 2007).

Amidst fears that a windfall tax would further frighten investors out of the country, the state backed away from implementing the tax. But the government's 2009 decision was seen as misguided when copper prices rapidly rose to reach all-time highs of $10,000 a tonne in the first half of 2011. Many advocated a reintroduction of the windfall tax. The rapid copper-fuelled growth led to an announcement in July 2011 of the country's fast journey to middle-income status that many viewed as a cynical political ploy for an election year. While it cannot be argued that some sectors of the population have benefited from this economic boost, with signs of increased consumption amongst a 're-emerging' middle class, much of the population still struggles to earn a living. In neighbourhoods of Copperbelt towns such as Luanshya, the landscape still displays the ravages of the withdrawal of the mines in infrastructural maintenance.

Present urban welfare on the Copperbelt

The previous section has provided an historical account of Zambia's heavy reliance on copper production, organisation of the copper industry, the capricious influence of the world copper price, and the rising success and later dismal plunge in urban welfare related to the state of the copper industry. The following section teases out the implications for the here and now, drawing attention to how the urban population's welfare and social identity are faring, beginning with the effects of privatisation on people's urban way of life.

Privatisation's impact on housing

The privatisation of state industry went hand in hand with the privatisation of urban housing, which had mainly been tied to employment from the colonial period onwards. The sale of housing to private individuals was enacted in 1997 by a presidential decree that stated that parastatal and council housing would be sold to 'sitting tenants'. This shift in policy had a huge impact on settlement in urban areas, and on the urban fabric of the Copperbelt. A formal job with the mines on the Copperbelt had previously meant a guarantee of shelter of a reasonable standard.

When the mines were privatised, beginning in 1997, the new owners were unwilling to take up these costly social responsibilities, and the administration of the mine townships was transferred in 2002 to the local authorities through a World Bank-funded programme. This meant that former mine employees, who had previously received free water and subsidised electricity, had to start paying bills. Many struggled to do so (Kazimbaya-Senkwe and Guy 2007). To meet the demand for water, residents of towns like Luanshya, who had been hardest hit by privatisation, initially drew water from an industrial supply that during the mining period had been used to water gardens. When this supply was closed, some dug wells in their backyards, and in the dry season, when the wells dried out, begged for water from neighbours who still had piped water. To meet their energy needs, many residents turned to charcoal-burning, chopping trees from the nearby forests and, more occasionally, their yards and the residential streets of the town.

A lack of maintenance of neighbourhood infrastructure was readily visible in pot-holed roads and the disappearance of storm water drains that had filled with silt over the years; darkened roads at night that had once been lit by street lights; burst water-pipes that it was speculated were vandalised to allow the illegal drawing of water; and sewer over-spills due to blockages, sometimes arising from an attempt to fertilise and water the acidic soils for commercial vegetable gardens. However, a garbage-disposal service continued, although, as many residents reported, there was little to dispose of, as poverty had ordained that what was discarded was useful for others.

Crime was a major concern with the end of the mine-policing system and poverty contributed to its increase. As the social reality on the Copperbelt changed, so too did residents' recreation activities. Grass sprouted in the cracks of the asphalt tennis courts of the mine recreation area in Luanshya. Fields of maize were planted here and there through the town. Anthills at the golf course were broken down to make building bricks. The local cinema was converted into a church since few could afford cinema-going. The squash club was converted into a drinking-place.

On the domestic front, residents who had previously competed to win a best-garden award, which the mines had put into place as an incentive to beautify the neighbourhood, no longer grew poinsettias, gardenias and other flower-bed plants. Nor did many bother to tend to their lawns, and dry swept-up patches of earth appeared around houses, reminiscent of the dry-earth surroundings of village homesteads. It was not only the impressions of the rural that gave rise to discussion of the Copperbelt becoming village-like; it was also the economic activities that residents engaged in.

While women's social life in particular bears some resemblance to aspects of village life, Copperbelt residents see themselves as modern. For this reason, they bemoan their towns that were 'beginning to look like villages', pointing not only to the towns' changing physical appearance, but to forms of sociality and mutual support that were suppressed or discarded when mine administrations dealt with town infrastructure and welfare needs. As residents adopted new forms of livelihoods, material landscapes were being reconfigured, infusing social relations with altered needs and meanings reminiscent of village life. This could be interpreted as going beyond Devisch's (1996) psycho-social notion of villagisation to encompass the material. To cite just a couple of concrete everyday examples, the tendency was manifested in more meals cooked on charcoal braziers and more water collected from wells or poached from neighbours. It also informed neighbourly interaction and rural-like forms of sociality, trespassing and conflicting with the Copperbelt's erstwhile nuclearising family life and residents' sense of modernity.

Informalisation of working lives: agriculture and small-scale mining to the fore

Before privatisation, the end of mine employment had necessitated moving out of the urban house and retiring to one's home village, a rural resettlement scheme or to some peri-urban area. With the sale of houses in the privatisation process, the retrenchees and retirees could opt to stay in their urban home, or, if not, generate a rental income from it (Mususa 2010a). Many Copperbelt residents gained a house as part of their retrenchment packages at the time of the worst social economic depression between 1997 and 2003. The house and yard became an important asset in sustaining urban livelihoods, with retrenched former miners staying in the lower-density former mine suburbs and carrying out agricultural activities that included growing vegetables, rearing chickens and pigs and fish farming. Residents of high-density suburbs with inadequate backyard space for sustainable backyard agricultural activity turned to farming on the periphery of the town.

In Luanshya mine township, following the loss of jobs, many residents started growing food in a future mine-development area. When plans got underway to develop this area in 2004, the new mining company evicted the farmers and built an electric fence around the perimeter, forcing the cultivators to move their fields further into rural Luanshya and Mpongwe, which entailed walking long distances and camping in the 'bush' during busy farming periods. Some residents decided to settle in these areas, and either sold or rented out their houses in town. These moves can explain why Luanshya's annual population growth rate between 2000 and 2010 was only 0.3%, while that of Mpongwe, the nearby farming block, grew by 3.6% (CSO 2011). This movement to the rural outskirts, and to forest reserve areas, resulted in several land rights problems (Hansangule, Feeney, and Palmer 1998).

Land contestation was not only confined to agriculturalists, but also to small-scale, mainly illegal, artisanal copper miners who emerged alongside the privatisation of the mines. Small-scale and illegal mining activity has contributed much to informal livelihoods of not only Copperbelt residents and those of the new mining areas of North Western Province, but also to those members of the Zambian public who were willing to migrate to those areas to exploit this opportunity (Mususa 2010b). On the 'old Copperbelt', women and children work on the copper-mine dumps, trespassing on the larger mine company's property to retrieve flux stone, a by-product of copper processing, which is then sold as building aggregate in the manufacture of concrete blocks. This industry has thrived in tandem with the rise of the small-scale building industry, spurred on by expanding home ownership and the construction of out-buildings for small livestock, as well as the maintenance of pot-holed roads.

In the 'new Copperbelt' in North Western Province, small-scale mining camps have emerged in the bush, either at older mine camps like that of Kalengwa mine, which was abandoned in the early 1980s, or, more commonly, in customary areas, under traditional authority. The populations of these camps have fluctuated since 2004 according to world copper prices. The majority of these camps have no basic services due to their rural location, but nonetheless, socially, the camps have an urban character that contrasts with the surrounding countryside (Mususa 2010b). They could also be described as 'rurban slums', as they accommodate a densely concentrated migrant population who are squatting on rural customary land.

Global mining boom, urban resurgence and the expanding poverty gap

With Zambia named amongst the 10 fastest growing world economies in the world between 2011 and 2015 by the *Economist* (6 January 2011), there is increasing pressure on the state to tax this copper-fuelled growth. The economic growth has sparked an increasing rate of urbanisation (CSO 2011). The urban/rural split of the population in the 2000 census was 35/65, moving to 39/61 in 2010. The average annual population growth rate in the 1990–2000 decade was 3.0% for rural and 1.5% for urban areas, contrasted with 2.4% for rural and 4.8% for urban areas during 2000–2010. On the Copperbelt, the drop had been more radical, with the decade 1980–1990 measuring an average annual growth rate of 1.9%, which dropped to 0.1% in 1990–2000, recovering to 2.2% during the years 2000–2010 as a result of the copper boom.

The region with the highest growth was Lusaka, which had an average annual growth rate of 4.7% during 2000–2010 (CSO 2011), reflecting its continued status as the country's premier city. Lusaka's growth does not, however, reflect a decent standard of life for the majority, but rather the fact that the city offers diversified and better opportunities and services. The minimum cost of living per month for a family of six in December 2010 for Lusaka was calculated at ZMK2,897,430. The average monthly income for a family in an urban low-cost settlement, which comprises 80% of all settlements in Lusaka, was ZMK645,326, while civil servants like teachers and nurses earned from ZMK1,300,000 to ZMK3,450,000 (JCTR 2010).

In contrast, in the rural district of Mufumbwe in North Western Province, where there is a combination of informal and formal copper mining activity, the residents of the area consumed on average 900 calories a day, a deficiency of more than half the 2400 calories required per day (JCTR 2009). In comparison with the Copperbelt and Lusaka, there is little social infrastructure, with residents having to travel on average more than 60 kilometres to access the mission hospital at Kasempa. With these inequalities and deficiencies in income, many Zambians feel they are not benefiting from the copper boom.

Foreign investment

Given the highly visible impact that changing foreign investment in copper has had on the welfare of the Zambian population, local perceptions and attitudes towards foreign investors in mining are continually expressed in people's everyday discourse on the Copperbelt. There is a general recognition of a major shift in responsibility for infrastructural and welfare provisions onto Copperbelt residents. A question often posed in local discussion is: 'what kind of investment is this, if they [foreign investors] can't build houses for us?', reflecting the still widely held expectation that mining investment will lead to modernisation for its employees.[4] Lumwana Mines plc, which is developing a greenfield site in the North Western Province, is held as a prime example of good investment because 'they are building houses'. The construction of housing for many Copperbelt residents is also seen as a gauge of mine-investors' commitment, showing their willingness to accept responsibility for the workers' welfare, as had been the norm in the years of mine-welfare largesse. In Luanshya, it was frequently commented that JW/Enya, the owners of the Luanshya mine at the beginning of the copper boom in 2004, had shown no signs of planning new housing. Apart from the rehabilitation of the public swimming pool and the former mine hospital, they did not undertake any major infrastructural development. The temporary nature of their stay was further highlighted by the conversion of offices into residences for some expatriate staff. When, in 2008, the mine was placed under care and maintenance, it confirmed the dim view of Luanshya residents on the short-term nature of their investment.

In 2009 the Luanshya mine was purchased by the China Non-Ferrous Metal Mining Company (CNMC). Despite general resentment towards Chinese investment on the Copperbelt, residents of the town expressed relief because this was perceived as longer-term largely due to significant investment in developing the infrastructure of the mine, and the well-publicised construction of what was to be the largest copper smelter in the Copperbelt. Others though were worried because there was no evidence that they were building any large-scale housing development and the Chinese management were staying in the old general manager's residence, which had been sub-divided into flats. One resident noted, 'they like to stay together', a practice confirmed by Lee (2010) in her study of Chinese enclaves in Zambia and Tanzania. This mode of living contrasted with that of former British managers who 'wanted swimming-pools immediately after they arrived'. In contrast, Lumwana Mines and First Quantum Minerals in the new mining area in North Western Province have planned golf courses.

The Chinese mining company's spartan approach to settlement, and a willingness to work long hours alongside the Zambians, are applauded as evidence of a strong work ethic. However, it is also problematic for Zambians because it alters the playing field for labour negotiations, which were previously conceptualised in terms of the moral inequalities embedded in the disparity of labour and living conditions between African and European mine workers, and later between African manual labourers as opposed to the managerial and political elite. According to Mr Hu, a Chinese citizen with business interests in Zambia, Zambians can no longer afford a work ethic and lifestyle inherited from the British; he extended his criticism further to observe that if Zambians learnt 'not to spend on Coca-Cola' their incomes would go further. Though President Kaunda had made a similar argument in the years of the copper decline, it is one that does not sit well with a majority of Zambians who point out that the majority are living below the poverty line, and struggle just to meet their basic needs.

Diversification from copper

Efforts are being made to attract new investment to diversify away from copper production that disproportionately contributes 11% to gross domestic product, and realised over 80% of the country's foreign exchange earnings in 2010. Regulations were put in place in 2009 to mine uranium in the North Western and Southern Provinces of the country. The viability of oil production and diamond mining is being explored in the Western part of the country that borders with Angola and Namibia. Gold is being mined by FQM at Kansanshi, but as a by-product of copper-mining operations.

The dramatic decline in living standards from the 1980s to the present day has renewed efforts both on the state level and the individual level to promote investment in agriculture. The Zambian state has been wooing displaced Zimbabwean commercial farmers and South African farmers worried about potential land redistribution to invest in the country. In the new mining areas of North Western Province, mines like First Quantum Minerals Ltd are beginning to experiment with supporting local out-grower schemes using high-yield techniques of conservation farming to supply the mines with fresh vegetables. Meat and grain production in close proximity to the mines is also encouraged to enhance food supply and provide alternative livelihoods for the long run.

At a household level, many Copperbelt residents carry out backyard and small-scale agricultural activities not only for subsistence but also as a way to earn an income (Mususa 2010a). Civil society, especially the Pentecostal churches preaching a prosperity gospel, has also been active in promoting farming as an alternative means of livelihood. One sermon entitled 'Getting back to God's Plan' – prepared by a Luanshya pastor (Kasonka 2008), prescribes a formula for creating a Garden of Eden, which would include livestock, birds and fruit, and result in a 'land flowing with milk and honey'. This kind of plan characterises the experiments of many Copperbelt residents in self-sufficient small-scale agriculture and livelihood diversification, which some would argue creates villagers in the town.

Conclusion: when the town becomes like a village

As the economy declined towards the end of the 1970s, the state increasingly called on people to become self-sufficient. On the Copperbelt, the mines in the 1980s promoted various schemes such as fish farming, sunflower cultivation and processing, poultry farming and similar activities. While the latter activities might not have been adopted widely, many Copperbelt residents recount going *kuma bala*, 'to the fields', in the 1980s, where they grew groundnuts, maize, pumpkins and sweet potatoes, mainly for home consumption. Most of these farming activities were taken up by mine employees' wives, their children and visiting relatives. The men, as several of my interviewees readily admitted, were to be found at the various recreation clubs that the mines offered while their relations farmed. Agricultural activities intensified in the more difficult years following the sale of the mines. A kind of pragmatism towards livelihoods set in as the number of waged labourers on the Copperbelt declined with massive retrenchments. With reduced income, going to rural areas became less viable as progressive years of reduced remittances from those in wage labour in the towns increased the harshness of rural poverty. In addition, the movement of rural kin to the towns reduced with the privatisation of the public transport sector and the previous benefits of a decent education in the towns became hard to support with the introduction of user fees.

The increase of urban poverty in the previously better-served Copperbelt towns can also be argued to have resulted in shifting exploitative capitalist relations from the mine to the home. Several incidents during the period of my research indicated that extended-family members, who by circumstances like the death of parents were living with former mine employees on the Copperbelt, were made to work long hours often without pay in the small enterprises that emerged. Neighbourly relations also became more fractious, as families without access to running water siphoned water from their neighbours. This usually occurred in the dry season when the wells they had dug in their backyards were completely dry. In retaliation, their neighbours closed off their outdoor water supply. Such difficult relations concerning basic needs suggest that the towns were increasingly like the village in material terms, but it was far from the ideal of convivial village life.

If this process of villagisation is conceived as people's pragmatic approach to realities of their changing material environment, one that encompasses multiple possibilities of dwelling and livelihood, then it is hard to see the Copperbelt as a place that is either urban or rural. Rather than dichotomising the Copperbelt – as urban or rural, modern or rustic, or viewed as in transition to becoming modern – it is best seen as textured by possibilities and constraints in an altered environment. In this way, places change not because of an abstract structure that stylises an idea of the modern or urban, but because our rooted entanglement with the environment grounds our experience and conceives of a reality that is readily discernible in what we do.

Globalisation, signalled by fluctuating world copper prices, is a grounded local experience for people living on the Zambian Copperbelt. Their experience is part of a network of inchoate to fully defined linkages between people and processes in the world at large. Zambian mining families' and policy-makers' multiple actions within that network transform the environment in ways that cannot be neatly contained. They encompass the lived effects of copper booms and busts, welfare approaches and abstract policies intended to stimulate economic growth. But definitely, the mines' withdrawal from the provision of welfare over the last two decades has been difficult and sometimes tragic for urban Zambians living on the Copperbelt, forcing them to re-evaluate who they are and what their future holds.

Acknowledgements

This article is part of PhD research supported by a Wadsworth African Fellowship (Wenner Gren Foundation for Anthropological Research) and the National Research Foundation of South Africa (Grant No. 63222, Ethnographies of the marginal. F. Ross, grantholder). The paper does not necessarily reflect the views of either the Wenner Gren Foundation or the NRF.

Notes

1. Bemba is the lingua franca of the Copperbelt.
2. Bank (2011) documents analogous contested place-making between urban modernity and rurality in the sub-urbanisation of Duncan Village in the South African city of East London.
3. A larger portion of this income was allocated to developing commercial farming and infrastructure for white settlement in Southern Rhodesia (now Zimbabwe) connected with the 1953 formation of the Central African Federation consisting of Nyasaland (Malawi) and Northern Rhodesia (Zambia).
4. N/Western Province elders urge FQM to build modern structures. *The Post*, 8 February 2011.

Notes on contributor

Patience Mususa is a PhD candidate in Social Anthropology at the University of Cape Town. She has published on women in the informal copper sector in Fraser and Larmer's 2010 edited collection and in Getting by: Life on the Copperbelt after the privatisation of the Zambia Consolidated Copper Mines. *Social Dynamics* 46, vol. 2 (2010): 380–94. Her research interests cover the social history of mining communities and, more broadly, the anthropology of the built environment. She can be contacted at: pmususa@gmail.com

References

Bank, L.J. 2011. *Home spaces, street styles: Contesting power and identity in a South African city*. London: Pluto Press.
Bissell, W.C. 2005. Engaging colonial nostalgia. *Cultural Anthropology* 20, no. 2: 215–48.
Bradley, K. 1952. *Copper venture: The discovery and development of Roan Antelope and Mufulira*. London: Mufulira Copper Mines Limited and Roan Antelope Copper Mines Limited.
Casey, E.S. 1997. Smooth spaces and rough edged places: The hidden history of place. *The Review of Metaphysics* 51, no. 2: 267–96.
Central Statistical Office (CSO). 2003. *Migration and urbanization 2000 census report*. Lusaka: Central Statistical Office.
Central Statistical Office (CSO). 2011. *Census of population and housing: Preliminary population figures*. Lusaka: Central Statistical Office.
Chauncey, G. 1981. The locus of reproduction: Women's labour in the Zambian Copperbelt, 1927–1953. *Journal of Southern African Studies* 7, no. 2: 135–64.
Coleman, F.L. 1971. *The Northern Rhodesian Copperbelt 1899–1962: Technological development up to the end of the Central African Federation*. Manchester: Manchester University Press.
Darby, C. 1931. Settlement in Northern Rhodesia. *Geographical Review* 21, no. 4: 559–73.
De Boeck, F. 1998. The rootedness of trees: Place and cultural and natural texture in rural Southwest Congo. In *Locality and belonging*, ed. N. Lovell, 25–52. London and New York: Routledge.
Devisch, R. 1996. Pillaging Jesus: Healing churches and the villagization of Kinshasa. *Africa* 66, no. 4: 555–86.
Epstein, A.L. 1981. *Urbanization and kinship: The domestic domain on the Copperbelt of Zambia 1950–1956*. London: Academic Press.

Epstein, A.L. 1992. *Scenes from African urban life: Collected Copperbelt essays.* Edinburgh: Edinburgh University Press.

Ferguson, J. 1999. *Expectations of modernity: Myths and meanings of urban life on the Zambian Copperbelt.* Berkeley: University of California Press.

Fraser, A. 2010. Boom and bust on the Zambian Copperbelt. In *Zambia, mining and neoliberalism: Boom and bust on the globalised Copperbelt,* ed. A. Fraser and M. Larmer, 1–30. New York: Palgrave Macmillan.

Fraser, A., and J. Lungu. 2007. *For whom the windfalls? Winners and losers in the privatization of Zambia's copper mines.* Lusaka: Civil Society Trade Network of Zambia.

Gewald, J., and S. Soeters. 2010. African miners and shape-shifting capital flight: The case of Luanshya/Baluba. In *Zambia, mining and neoliberalism: Boom and bust on the globalized Copperbelt,* ed. A. Fraser and M. Larmer, 155–84. New York: Palgrave Macmillan.

Haglund, D. 2010. From boom to bust: Diversity and regulation in Zambia's privatized copper sector. In *Zambia, mining and neoliberalism: Boom and bust on the globalized Copperbelt,* ed. A. Fraser and M. Larmer, 91–126. New York: Palgrave Macmillan.

Hansangule, M., P. Feeney, and R. Palmer. 1998. *Report on land tenure and insecurity on the Zambian Copperbelt.* Zambia: Oxfam Great Britain in Zambia.

Heisler, H. 1971. The creation of a stabilized urban society: A turning point in the development of Northern Rhodesia/Zambia. *African Affairs* 70, no. 279: 125–45.

Jesuit Centre for Theological Reflection (JCTR). 2009. *Mufumbwe, basic needs basket.* October. Lusaka: JCTR.

Jesuit Centre for Theological Reflection (JCTR). 2010. *Lusaka, basic needs basket.* December. Lusaka: JCTR.

Kasonka, E. 2008. *Getting back to God's plan for man.* Luanshya: Self published.

Kaunda, K., and C. Morris. 1966. *A humanist in Africa.* London: Longmans, Green.

Kazimbaya-Senkwe, B., and C.S. Guy. 2007. Back to the future: Privatization and the domestication of water in the Copperbelt Province of Zambia, 1900–2000. *Geoforum* 38, no. 5: 869–85.

Larmer, M. 2007. *Mineworkers in Zambia: Labour and political change in post-colonial Africa.* London and New York: Taurus Academic Press.

Larmer, M., and A. Fraser. 2007. Of cabbages and King Cobra: Populist politics and Zambia's 2006 election. *African Affairs* 106, no. 425: 611–37.

Laterza, V. 2008. Lost opportunity for democracy in Zambia. *Mail and Guardian,* November 5.

Lee, C.K. 2010. Raw encounters: Chinese managers, African workers and the politics of casualization in Africa's Chinese enclaves. In *Zambia, mining and neoliberalism: Boom and bust on the globalized Copperbelt,* ed. A. Fraser and M. Larmer, 127–54. New York: Palgrave Macmillan.

Lungu, J. 2008. The politics of reforming Zambia's tax regime. *Southern African Resource Watch* 8: 2–28.

Lusaka Times. 2010. Zambia: Those were the days!, March 26.

Mitchell, C. 1954. African urbanization in Ndola and Luanshya. *Rhodes Livingstone Communication* 6. Lusaka: Rhodes Livingstone Institute.

Mususa, P. 2010a. Getting by: Life on the Copperbelt after the privatization of the Zambia Consolidated Copper Mines. *Social Dynamics* 36, no. 2: 380–94.

Mususa, P. 2010b. Contesting illegality: Women in the informal copper business. In *Zambia, mining and neoliberalism: Boom and bust on the globalized Copperbelt,* ed. A. Fraser and M. Larmer, 185–208. New York: Palgrave Macmillan.

Mutale, E. 2004. *The management of urban development in Zambia: International land management series.* Aldershot: Ashgate.

Myers, G.A. 2003. *Verandas of power: Colonialism and space in urban Africa.* Syracuse, NY: Syracuse University Press.

Potts, D. 1995. Shall we go home? Increasing urban poverty in African cities and migration processes. *The Geographical Journal* 161, no. 3: 245–64.

Powdermaker, H. 1962. *Copper town: Changing Africa, the human situation on the Rhodesian Copperbelt.* New York: Harper and Row.

Prain, R.L. 1956. The stabilization of labour on the Rhodesian Copperbelt. *African Affairs* 55, no. 221: 305–12.

Schumaker, L. 2008. Slimes and death dealing dambos: Water, industry and the Garden City on Zambia's Copperbelt. *Journal of Southern African Studies* 34, no. 4: 823–40.

The Economist. 2011. Africa's impressive growth. January 6.

The Post. 2011. N/Western Province elders urge FQM to build modern structures. February 8.

The Post. 2011. Mopani pilot audit reveals tax payment irregularities. February 9.

Van Binsbergen, W. 1998. Globalization and virtuality: Analytical problems posed by the contemporary transformation of African societies. *Development and Change* 29: 873–903.

Van Donge, J.K. 2008. The plundering of Zambian resources by Frederick Chiluba and his Friends: A case study of the interaction between national politics and the international drive towards good governance. *African Affairs* 180, no. 430: 69–90.

Wilson, G. 1968. *An essay on the economics of detribalization*. Manchester: Manchester University Press for the Rhodes-Livingstone Institute.

Of prosperity, ghost towns and havens: mining and urbanisation in Zimbabwe

Amin Y. Kamete

Mining has always played a significant role in the urbanisation of Zimbabwe. That Zimbabwe's most urbanised province has the highest concentration of mining settlements is no accident. Whereas mining initially gave impetus to urbanisation, economic growth and prosperity, its decline precipitated urban shrinkage and poverty. In the wake of the post-2000 crises, shrinking mining towns have become havens. This discussion unravels the nexus of mining, urbanisation, governance and poverty. Using the case of chromite and copper mining, the article examines how mining, initially a driver of urbanisation and prosperity, later became the catalyst for decline and impoverishment. It argues that the fate of mining towns cannot be explained without reference to the intersection of national governance, local conditions and globalisation. The article makes two central observations: mining settlements are an integral part of the urban hierarchy; and the resurgence of mining towns is more of a mutation than a revival.

With under 40% of its population classified as 'urban' in 2010 (UNDESA 2007), Zimbabwe is not a highly urbanised country by world or sub-Saharan African standards. This characterisation is especially valid when the absence of mega-cities in Zimbabwe is taken into account, what with Harare, the capital and primate city, having a population of under 1.6 million (UNDESA 2007). Even so, Zimbabwe has experienced pulses of rapid urbanisation and shrinkage, as well as industrialisation and de-industrialisation. These pulses have been mediated by Zimbabwe's contentious economic and political trajectory.

In this, mining has played a crucial, albeit unheralded, role. It was mineral wealth that prompted the colonisation of the country; it was mining that was instrumental in spurring urban growth; and it was the closure of mines that heralded urban shrinkage. In what follows, I reflect on this unexplored facet of urban Zimbabwe. Using one province, I examine the nexus of mining, urbanisation, governance, prosperity and poverty. I show that while mining was a driver of urbanisation, economic growth and prosperity in colonial Zimbabwe and in the first decade of independence, it became the chief agent of decline, thanks to a combination of plummeting world mineral prices and ruinous government policies. I also discuss how, in the wake of the post-2000 crises, the declining mining towns have become havens for internally displaced persons (IDPs).

After the introduction, I present a brief conceptual overview on mining and urbanisation. I then go on to present a history of mining and urbanisation in Zimbabwe. This is followed by a snapshot of mining and urbanisation in Mashonaland West Province. I then analyse the mining–urbanisation couplet, focusing on chromite and copper. Finally, before the conclusion, I critically reflect on the impact of national governance on mining and urbanisation.

Mining and urbanisation

It has long been recognised in settlement analysis that one reason for the genesis of human settlements that sometimes grow into cities is the location of resources, including minerals (Knox and McCarthy 2005, 71). For mining towns, the initial propellant, and therefore the 'economic base', is constituted by mining operations. Here, mining constitutes the basic sector, which exports and brings wealth from outside, with everything else, including social and physical infrastructure and services, constituting the 'non-basic sector' (Brail, Bossard, and Klosterman 1993). Mining towns are engines of capital accumulation (Amin 2008); their fortunes – boom, bust, prosperity and poverty – are intertwined with the initial propellant and the basic sector, namely mining.

In this compelling, and rather mechanistic analysis, single-resource mining towns will grow and prosper as long as mineral extraction remains viable and the site maintains its competitive advantage (Leadbeater 2009). With its immobile resource, the mining town can diversify, becoming the source of agglomeration economies (Bailly, Jensen-Butler, and Leontidou 1996). However, once the basic sector ceases to exist or becomes unviable, the town no longer serves its purpose as a centre of capital accumulation. It is in danger of being abandoned and becoming a ghost town. Its resilience or vulnerability is dependent on the level of functional diversification. Hence, the greater the level of diversification, the greater the capacity of the town to 'restructure' and adapt to the new imperatives (Logan and Swanstrom 2005).

While this economistic and apolitical model is compelling in its simplicity, its applicability to contemporary Africa is problematic, thanks to a key omission: national governance. The trajectory of the African mining town is an outcome of complex local and global economic forces; but these forces are often mediated by state policies, themselves a product of national governance (Weiss 1999). As the case of Zimbabwe demonstrates, the ebbs and flows may not reflect the logic of local, national and global economic realities, thanks to government policy, action and behaviour. This is not unique to Zimbabwe. In a study of Zambia's Copperbelt – the site of quintessential mining towns in Africa – Mutale (2004, 67) identifies 'political and economic structures...[and] power relationships' that have influenced the development of the mining towns. National governance does generate certain imperatives capable of spawning social, political and humanitarian situations that ultimately determine the fate of mining-dependent towns, sometimes in defiance of local and global economic logic. As Molotch (1976, 312) noted, government decisions impact on local corporations and 'help determine the cost of access to markets and raw materials, ... influence the cost of overhead expenses ... [and] affect the costs of labour'.

National governance can erode the competitiveness of the mining industry, thereby adversely affecting mining towns in particular and urbanisation in general.

Government policies that contradict the logic of economic globalisation include trade, foreign exchange, labour and price controls (Bond and Manyanya 2002). These policies are often politically motivated, hence their short-sightedness. At the same time, policies that are 'globalisation-friendly', such as trade liberalisation and privatisation, can wreak havoc on the local mining industry, with direct consequences for mining towns (Leadbeater 2009). As the case of Zimbabwe shows, sometimes the government acts in overtly political ways with unintended consequences on the economy and settlement system, including the urban hierarchy. Political repression and politically motivated violence, for example, have been known to result in dislocation and displacement, giving rise to large numbers of IDPs (Lanjouw, Mortimer, and Bamforth 2000). The same applies to economically ruinous policies that destroy livelihoods, forcing people to look or relocate elsewhere for survival (Brand 2001; Kamete 2012). In Zimbabwe the havens that have offered refuge and 'replacement livelihoods' are the neglected and forgotten mining settlements.

The mining industry in most of Africa is export-oriented. Consequently, mining towns are at the mercy of global market forces, which ultimately determine world mineral prices. Globalisation has meant that 'governments have mostly lost control of the national economy' (Weiss 1999, 60). It is global forces that influence profitability and risk, thereby determining two important variables in the mining industry: investment and production. In turn, investment and production inevitably affect the growth or shrinkage of mining settlements whose fortunes are entangled with mining operations. How well the mining industry – and ultimately mining towns – copes is a function of the interaction between national governance and global market forces.

Mining and urbanisation in Zimbabwe

Historical overview

Minerals were crucial to the colonisation of Zimbabwe by the British in the late nineteenth century. Settlers in the British South Africa Company (BSAC) expected to find a new Eldorado in Mashonaland (Hensman 2007). In fact, the granting of the Royal Charter to Cecil John Rhodes' BSAC was made possible by the presentation of a mining concession Rhodes had obtained from King Lobengula of Matebeleland (Parsons 1993). From the outset, as the BSAC set about colonising Mashonaland, emphasis was placed on control over precious metals and other mineral resources (Bryce 1970). Not surprisingly, mining was to play a significant role in the history of the settler community and the spatial development of the colony, a feature that was to spill over into post-colonial Zimbabwe.

In the process, mineral resources were exploited and mining settlements developed. Because the exploitation of mineral resources was entrenched in private capital, with the state's role being limited to creating the regulatory framework and marketing, the sector experienced cycles of boom and bust. These pulses were closely linked to the national economy; they were also influenced by national political and economic governance. Additionally, at some key moments, they were also strongly mediated by international markets. The trend continued into the twenty-first century. By then, Zimbabwe had become a major mining country. As Saunders (2008, 68) observes, in the last decade of the twentieth century 'Zimbabwe was poised to become a significant force in African mining'.

Despite the biting economic crisis of the first decade of the twenty-first century, the embattled government recognised the important role of mining in the revival of the battered economy. At the height of the economic crisis in 2005, the Governor of the Reserve Bank of Zimbabwe (RBZ) acknowledged that the central bank was 'cognisant of the mining industry's importance in generating foreign exchange, employment as well as the linkages with manufacturing [sic]' (Gono 2005, 51). This was despite the fact that in 2004 'the mining sector [had] underperformed by 13.8%' (Gono 2005, 8). Predictably, 'increased mining production' was one of the goals of the ill-fated National Economic Recovery Programme (NERP) (Zimbabwe 2004, 16). This emphasis is not misplaced; mining has played, and has the potential to play, a crucial role in Zimbabwe's economic and spatial development.

Mining and urbanisation

Table 1 provides a snapshot of mineral exploitation in Zimbabwe. The table captures only large-scale formal mining and excludes numerous artisanal chromite mining ventures along the Great Dyke. As the table shows, even when focusing exclusively on formal mining ventures, with more than 80 operating mines of various sizes scattered throughout the country, Zimbabwe is a mining country. Unsurprisingly, mining has had a huge influence in the development of human settlements.

Zimbabwe's mining attracted major multinationals. They include the UK's Lonrho and Rio Tinto; South Africa's Anglo American, Anglo Platinum, Mettalon and Implats; Australia's Broken Hill Proprietary Company (BHP) and Delta Gold; and Canada's Casmyn Corporation and Trillion Resources. With such wealth and investment in minerals, Zimbabwe, perhaps inevitably, experienced significant mining-led urbanisation.

In this analysis the conceptualisation of the 'urban' follows official definitions. The Central Statistical Office (CSO 1993) has adopted a multidimensional approach to the designation of urban areas. For any settlement to pass the 'urban test' it should (1) have been administratively declared as an urban area; (2) have a population of at least 2,500; (3) have a compact population pattern; and (4) have the 'majority' of its workforce engaged in non-agricultural activities. The difference between a 'city' and

Table 1. Mineral exploitation in Zimbabwe (2010).

Mineral	Number of mines
Asbestos	3
Coal	3
Copper	2
Diamond	2
Gold	61
Iron ore	2
Nickel	6
Platinum	1
Chromite*	5

Source: All except * based on MBendi (2011); * ZIMASCO (2010).

any other town is purely one of administrative definition. A city is declared to be so by central government; a town is any urban centre that is not a city.

Urban centres that originated from and/or depend on mining are shown in Table 2. Two cities owe their origins and existence to mineral extraction and/or mineral processing. Kadoma, with a population of some 76,000 (World Gazetteer 2011), is located in a mineral-rich area that is home to gold, copper and nickel mines. Kwekwe, which has a population of about 100,000 (World Gazetteer 2011) is Zimbabwe's centre of iron and steel and fertiliser manufacturing. Apart from these two cities, there are numerous other major urban centres that originated from mining, whose growth was driven by mining and whose prosperity depends on mining. The midlands town of Redcliff (population 34,000) owes its existence to the Zimbabwe Iron and Steel Company's (ZISCO) iron and steel works. Hwange (population 34,000), in Matebeleland North, is a coal-mining town whose fortunes are intertwined with the colliery.

Table 2 shows only those urban centres that meet the criteria that a settlement should meet 'to pass the urban test' (Kamete, Tostensen, and Tvedten 2001, 5). The centres in Table 2 pass the 'urban' test. They have a combined population of some 423,000. This constitutes slightly over 8% of the national urban population of 5,264,000 (UNDESA 2007). It should be noted that the table does not include numerous smaller towns that do not administratively qualify as urban centres and a lot of 'fallen' ghost towns that collapsed following the closure of mines in the 1980s and 1990s.

Table 2. Mining towns in Zimbabwe (2011).

Centre	Province	Population[1]	Mineral extraction/processing
Kwekwe	Midlands	99,578	Iron & steel works; fertiliser
Kadoma	Mashonaland West	77,498	Gold, copper, and nickel mining
Chegutu	Mashonaland West	46,366	Gold
Zvishavane	Midlands	34,855	Asbestos (also gold, platinum, beryl)
Redcliff	Midlands	34,607	Iron and steel works
Hwange	Matabeleland North	34,187	Coal; thermal power generation
Shurugwi	Midlands	16,697	Chromite, gold and nickel
Mashava	Masvingo	12,277	Asbestos
Inyati	Manicaland	8,633	Gold and copper
Trojan Mine	Mashonaland Central	7,794	Nickel
Penhalonga	Manicaland	7,546	Gold
Mhangura	Mashonaland West	7,517	Copper
Mutorashanga	Mashonaland West	7,189[2]	Chrome
Chakari	Mashonaland West	6,465	Gold
Alaska Mine	Mashonaland West	4,745	Copper mining, smelting and refining
Arcturus	Mashonaland East	3,866	Gold
Renco	Masvingo	3,708	Gold
Patchway	Mashonaland West	3,590	Gold
How Mine	Matabeleland North	3,287	Gold
Brompton	Mashonaland West	2,725	Gold

Notes:
[1] All population figures, except Mutorashanga, are 2010 estimates from World Gazetteer (2011).
[2] Based on 2002 census (CSO 2002).

However, this figure should be placed in context. Though some areas started off as mining settlements or as homes to mineral processing works, they did assume a life of their own as other activities, especially in the services and retail sector sprang up to capitalise on the backward and forward linkages arising from mining operations. Kwekwe, Kadoma and Chegutu are good examples. Similarly, other urban centres that do not qualify as 'mining towns' have at one time depended, or still depend, on the surrounding mining or mining-related activities for growth and prosperity. Chinhoyi, Marondera, Gweru and Bindura – all of them provincial capitals – fall into this category.

The capital city, Harare, has not been immune to the influence of mining. Most of the national headquarters of some blue chip national and multinational companies engaged in mining are located in the city. In addition, many of the manufacturing and service industries in Harare and Bulawayo, the second largest city, are linked to the mining industry in the form of forward and backward linkages; so are some specialised service companies and trading outlets. By extension, this means that a lot of the employment in many sectors in the cities is linked to mining. The same can be said of the attendant housing. In view of the foregoing, it can be concluded that even these large cities owe some of their growth and fortunes to the mining industry.

Chromite and copper mining settlements in West Mashonaland

Mashonaland West is one of the 10 administrative provinces of Zimbabwe. Popularly known as 'Mash-West', the province has an area of 57,441 square kilometres and a population of approximately 1.2 million according to the 2002 national census (CSO 2002). The provincial capital is Chinhoyi with an estimated 2010 population of 61,867 (World Gazetteer 2011). There are six administrative districts in Mashonaland West, namely Kariba, Hurungwe, Zvimba, Makonde, Kadoma and Chegutu (Figure 1). Mashonaland West is rich in natural resources. Lying mostly in Agricultural Region 2, with good soils and rains, the province was, until the decimation of commercial agriculture (see below), home to the country's most productive large-scale commercial farms. It is still the breadbasket of Zimbabwe.

With respect to mining, the province boasts rich mineral reserves and some of Zimbabwe's biggest mines and mining towns. Some notable mining towns include Kadoma, Mutorashanga, Mhangura and Alaska Mine (see Table 2). Chromite, copper, gold, nickel and platinum group metals are some of the province's mineral resources that have been exploited. These minerals have given rise to some of the province's large mining settlements and urban centres.

What follows will discuss the fortunes of towns that depended on chromite and copper. It does not pretend to offer a comprehensive history or inventory of mineral resources. Rather, it is intended to provide insight into the developments that shaped the rise, fall and mutation of mining towns with a view to, first, linking mining and urbanisation in the province and, second, explaining why the province is where it is today in terms of mining, urbanisation and poverty.

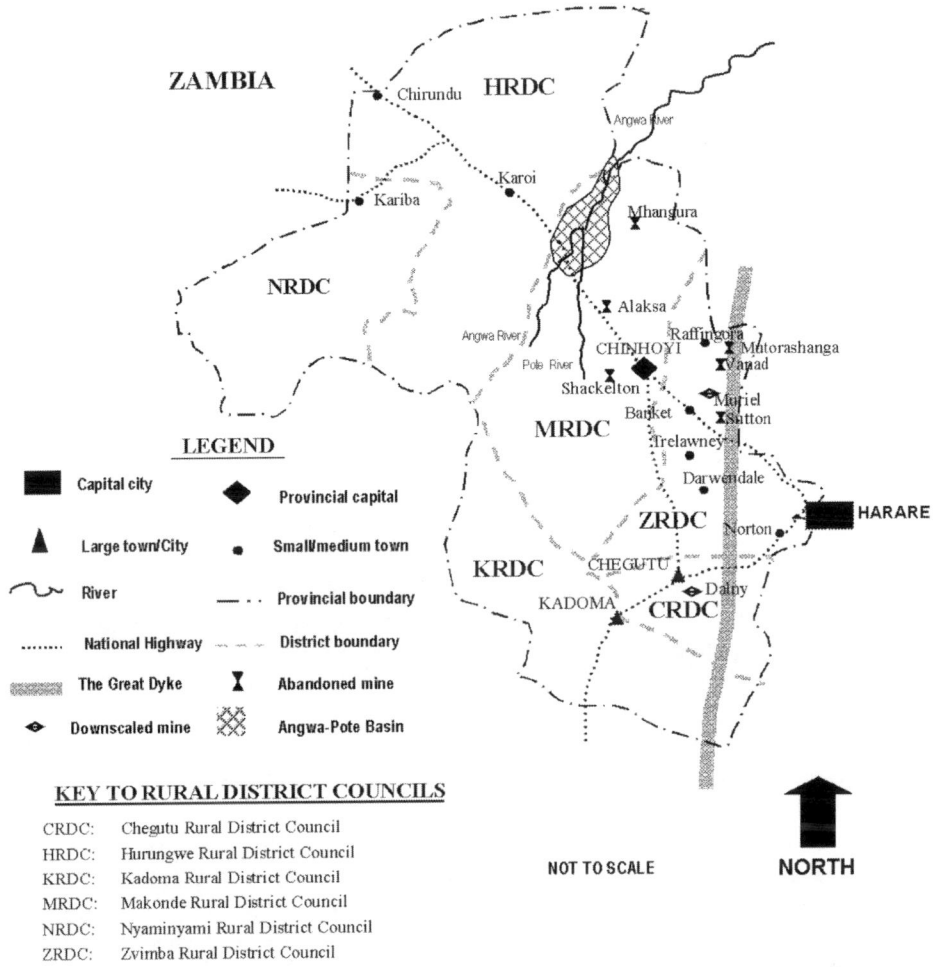

Figure 1. Districts of Mashonaland West (Kamete 2008).

Copper mining

Mashonaland West was home to Zimbabwe's largest copper mines: Mhangura, Shackleton, Alaska and Sanyati. Sanyati Mine in Kadoma District is no more than a village, so this paper will focus on the first three mining settlements situated in what is now Makonde Rural District (MRD). Shackleton was a copper mine of some 5,000 people. Alaska was slightly more diversified. In addition to copper mining, it had copper smelting and refining facilities. At its peak, the settlement had a population of some 7,000. Mhangura, the largest of the copper mines had a population of around 15,000 (CSO 1993); it also had a copper refinery and smelting plant. Mining at both Alaska and Mhangura started in 1959.

Of the three mining settlements, Mhangura blossomed into something resembling a 'real' town (see below). Sprawling and segregated residential areas were complemented by some form of a Central Business District (CBD), commercial banks, educational and health facilities and administrative and service functions including a post office and police station. The town became known for its good sports facilities and success in sports. With its smelting and refining works Alaska also blossomed, though not to the level of Mhangura. Shackleton was mostly residential with a spattering of ancillary services; it cannot be said to have become a successful town. Most of the people employed at the mines were migrants from Malawi, Zambia and Mozambique.

The three settlements were true mining towns. They owed their origin and existence to copper mining and/or processing. Not surprisingly, their destiny was intertwined with that of copper mining. In the mid-to-late 1990s, when world mineral prices collapsed and production became unviable, the mining ventures suffered viability problems. To cope, the mining companies initially scaled down operations. When the problems persisted, they collapsed and closed down. Mhangura, which had been taken over – and systematically mismanaged (Moyo 2010) – by the parastatal Zimbabwe Mining Development Corporation (ZMDC), was the last one to close down in November 2000 due to bankruptcy (Sango et al. 2006). With mining no longer able to buoy the local economy, the once bustling settlements spiralled into seemingly terminal decline. Some migrants were repatriated with the help of their employers. Shackleton, which had not managed to diversify like Mhangura and Alaska, was the least resilient. It was virtually abandoned (IPA-SC 2001). By 2007 Zimbabwe was producing a paltry 3,000 metric tonnes (mT) of copper compared to 19,200 mT in 1994 (Joseph and Kundig 1998). As a result, Alaska, which still retains its smelter and refinery, now handles mainly imported concentrates.

Having lost their source of livelihood, most of the residents of the urban centres descended into poverty (Zimbizi 2001). In a way this was to be expected. The bulk of the workers were unskilled labourers who had known no other job; some were second-generation miners. At the time, other mines – especially chrome and gold mines – were experiencing a swift downturn (Maponga and Ruzive 2002). The mining labourers' lack of training in other trades did not help. Besides, at the time jobs were scarce in Zimbabwe. Thanks to the Economic Structural Adjustment Programme (ESAP), there were mass retrenchments everywhere, especially in mining and manufacturing (Bond and Manyanya 2002). That most of the labourers were migrants exacerbated their woes. They had no rural home to fall back on (Potts 1995), and therefore no land for housing or livelihood. By 2000, Zimbabwe was no longer a copper mining giant. Shackleton was a ghost town; Mhangura and Alaska had shrunk.

Chromite mining

It is estimated that Zimbabwe, which accounts for 3% of world chromite production, holds 21% of the world's chromite reserves (Mobbs 2007). In 2007, Zimbabwe produced 650,000 tonnes of chromite ore (Mobbs 2007). About 8,000 people were employed in officially registered chromite mines, accounting for 13% of the labour force in the mining industry (Maponga and Ruzive 2002). This figure applies to the period when chromite mines had collapsed and the economic crisis was beginning to bite. In its heyday, chromite mining was a huge industry in terms of both production and employment. The inclusion of unregistered mines, which are littered throughout the Great Dyke, would boost the employment figures in the chromite mining industry.

Chromite mining spans the entire mineral-rich Great Dyke (Figure 1). Most of the mining claims are owned by two multinational corporations: Zimbabwe Mining and Smelting Company (ZIMASCO), formerly a subsidiary of Union Carbide, and now a subsidiary of China's Sinosteel Corporation, and Zimbabwe Alloys (Zimalloys), a subsidiary of the Anglo American Corporation. In Mashonaland West, chromite mining takes place on the northern sections of the Great Dyke. It is located predominantly in Zvimba Rural District (ZRD), which is under the jurisdiction of Zvimba Rural District Council (ZRDC).

The growth of mining settlements was a direct result of chromite mining activities during the colonial era. Several settlements were developed when mineral deposits began to be exploited. Apart from a few mines such as the Lonrho-owned Muriel gold mine, most of the mines were based on the extraction of chromite. As was the case in all towns, the development of the mining settlements was mediated by the racist ideology of the colonial administration. Under the Land Apportionment Act (1930), which divided the country into black and white areas, the chromite mines were located on land classified as crown (state) land; the surrounding farming areas were classified as 'white' areas. Blacks could not own property in these areas. They could only settle as paid labourers. Like the situation in the copper mines, most of the mine labourers were migrants from Malawi, Zambia and Mozambique. The locals skirted mining work because of low wages and poor working conditions (Save the Children 2003, 8).

At independence in 1980, there were several established chromite mining settlements that passed the urban test. Mutorashanga, the largest, was perhaps the most 'conventional' of the towns. It had a diversity of functions and a CBD. The other towns were Sutton, Vanad and Kildonan. In time, after independence, informal settlements sprang up around the mining towns; they include Chrome, Kimco, Tavakuenda and Shunguyaguma. According to the 2002 census, together these towns, excluding the informal settlements, had a population of about 9,000 (CSO 2003). If we were to include the informal settlements and Caesar Mine – which, though on the Mashonaland East side of the Great Dyke, has more links with Mash-West than its own province – this would increase the population to between 13,000 and 15,000 (Moyo 2010).

Just like the copper mines, the settlements reflected the centrality of mining. Housing dominated the settlements, with services and community facilities that supported and/or depended on the primary mining activities constituting the land use types at the mining towns. Like Shackleton, Sutton, Vanad and Kildonan did not diversify much, though Sutton seemed to have done better with a milling plant and a bustling township centre that could rival any in the more conventional towns like Banket. This lack of diversification meant that the single-resource towns were vulnerable. Any shock that hit chromite mining was certainly going to hit them hard.

That shock came in the mid-1980s. The decline in world mineral prices, coupled with government policy made mining unviable (Save the Children 2003; Bond and Manyanya 2002). This resulted in the two companies (ZIMASCO and Zimbabwe Alloys) resorting to mass retrenchments, terminating contracts of part-time labourers and scaling down to remain afloat (Maponga and Ruzive 2002). In the end, still dogged by viability problems, the mining companies wound down their operations. The state-owned ZMDC stepped in and organised the retrenched workers into chromite mining cooperatives. Initially, ZMDC hired out equipment and provided managerial advice (Maponga and Ruzive 2002). The cooperatives were allowed to work on ZIMASCO and Zimalloys claims. They would produce chromite on contract and were paid according to tonnage and product quality.

In a way, the demise of the chromite mines heralded a moment of urban shrinkage. In an unprecedented repatriation programme, many migrant workers were assisted to return to their countries of origin by their erstwhile employers. Some skilled workers relocated to other mines. They left behind large numbers of empty houses. The smaller towns of Sutton, Vanad and Kildonan resembled ghost towns, though not to the same level as Shackleton. Though severely depopulated (one mining foreman insisted 50% of the population had left) (Mwanza 2010) the towns were not completely abandoned. Life went on – though without the vibrancy the settlements had experienced when mining was still alive – sustained by the tribute mining and other 'off-mine' activities (Kamete 2008). The centres still had functioning schools, shops and services such as primary health care and public transport.

Currently, apart from the resuscitated but scaled-down ZIMASCO underground mine at Mutorashanga, all mining in the area can be classified as 'tribute mining', where 'a tribute is a lease of a registered mining claim by the claim owner to another' (Save the Children 2003, 8). Under the system, mining is undertaken by individuals, cooperatives and small and medium mining enterprises which act as tributors to ZIMASCO and Zimalloys. Most of the annual production is done by tributors. In 2007 they produced 230,000 mT compared to ZIMASCO's mines which together produced a total of 120,000 mT (Mobbs 2010).[1] The ore is transported to Gweru and Kwekwe for industrial use.

A resurgence of sorts: from shrinking towns to havens

The copper and chromite mine towns in Mashonaland West experienced some kind of resurgence at the beginning of the twenty-first century when they became havens. In this paper a haven refers to a safe or peaceful place that offers shelter, safety or retreat; it is a refuge. This was due mainly to two controversial national developments: a chaotic land reform programme and a cataclysmic urban clean-up campaign. Both these epochal events had significant impacts on mining settlements. This is mainly because they were characterised by some common traits and outcomes, chief among them being violence, displacement and dislocation.

Land reform

In 2000, the government embarked on the ruinous Fast Track Land Resettlement Programme (FTLRP). The programme has been credited with two distinct deleterious outcomes on the national economy with corresponding damage to the social fabric. The first one is the destruction of the large-scale commercial agricultural sector. The decimation of the sector resulted in the displacement of over 300,000 farm workers and their families (Sachikonye 2003; Hartnack 2005). The dislocated farming communities ended up without livelihoods, and, for some of them, without homes. The second outcome was the virtual decimation of commerce and industry. Particularly adversely affected were those sectors with extensive backward and forward linkages to commercial agriculture (Gono 2005; IMF 2005). This is attributable to the drastic curtailment of activities on commercial farms, which occurred within months of the eruption of land occupations (Fewsnet 2002). Many of the firms in the agro-industrial sector and related service industries lost markets and/ or service providers and/or input suppliers (IMF 2005).

Many of the displaced farm workers and retrenched urbanites trekked to the mines. Perhaps it was to be expected. Without houses to their name, and unable to afford the urban housing rentals – which were constantly being hiked because of spiralling inflation – the displacees and/or jobless sought refuge in the lethargic mining towns which had an abundance of vacant housing. They had no option. As stated above, most of them being foreigners, they had no rural home to go back to. It was easier and cheaper to relocate to the mining settlements where decent housing could be acquired at very reasonable rentals – the landlords being the local authority who took over the administration of the settlements from the mining companies. Here, the displacees and the jobless could secure accommodation and earn a living.

Those who chose to settle in the conventional towns ended up renting substandard housing in informal settlements or in backyard shacks in the high-density residential areas. They joined the ranks of informal sector practitioners or took up low-paid menial jobs. Others joined the 'new gold rush' and became gold panners (Kamete 2008). A few took up cross-border trading, while others went into a life of crime. Within a few years, those who had taken refuge in the informal sector were to experience another calamity (see below).

With so many people flocking to the mining towns, in a way the centres experienced some kind of resurgence – at least in terms of population. According to a relief worker, the population in the areas surged by more than 60% (Mrs LM 2010). In the copper towns, the population, which had fallen to below 20,000, shot above 30,000 (see IPA-SC 2001). In the chrome towns the population now hovered around 15,000 compared to a low of 8,000 (Zimbizi 2001; Mrs LM 2010).

The urban clean-up campaign

'Operation *Murambatsvina*/Restore Order' (OM/RO) is the official designation for Zimbabwe's urban 'clean-up' campaign that officially ran from May to July 2005. Government embarked on this blitz ostensibly to 'reject and eject filth' (Kamete 2007). As the campaign got underway, the identity of the 'filth' became apparent. The human part of this 'filth' comprised, predominantly poor urban residents (Potts 2006). Among them were the homeless (vagrants and street children), vendors operating in undesignated places, and lodgers (informal tenants) housed in unauthorised structures in informal settlements or in the high-density, low-income residential areas. The police commissioner described this human 'filth' as a 'crawling mass of maggots bent on destroying the economy' (*The Herald* 2005). The non-human part of the 'filth' consisted of unauthorised settlements, illegal residential and informal business structures, as well as animals and equipment kept in residential and other premises in contravention of a cocktail of legal and regulatory controls (Kamete 2009).

During the blitz – nicknamed 'The Tsunami' by Zimbabweans – the army, police and youth militia carried out three related activities: demolition, eviction and detention/relocation. They demolished illegal residential and business structures in informal settlements and low-income residential areas; they forcibly and violently evicted vagrants, street children and vendors operating in undesignated places; and they arrested, detained and/or forcibly relocated the human 'filth' to 'transit camps' or 'holding centres' where they were to be vetted so that the authorities could determine what to do with them (SPT 2005; ICG 2005).

The systematic bulldozing and destruction of structures, mostly in the poorest parts of Zimbabwe's urban areas, resulted in some 700,000 urban dwellers losing homes and/or livelihoods (Tibaijuka 2005). By mid-July, 92,460 residential structures had been destroyed, negatively affecting 133,534 families; this in addition to 32,538 business structures razed to the ground, and 17,512 households that were officially 'unaccounted for' (Tibaijuka 2005, 85).

The victims of OM/RO had limited choices. Among the 'most viable' were two: to go back to their rural homes or to go to the mining towns, which had not been touched by the Tsunami. Just as was the case in the FTLP, most of the victims were foreigners with no rural home. In any case, many of the victims of the clean-up campaign were already victims of the FTLRP. In the town of Banket alone, more than three in four of the victims of OM/RO had come from the farms (Mrs LM 2010). Not surprisingly, a substantial proportion of the displaced urbanites trekked to the mines. There are no figures to quantify the new arrivals; a civil society activist who works in the chrome and copper mines claims that 'the influx was very, very phenomenal' (Moyo 2010).

Havens – but not 'real' towns?

Following the two waves of displacement, dislocation and violence, the mining settlements were nowhere near the ghost town status they had acquired. They were back in business. But life was not as at it had been before the large-scale mines were abandoned and tribute mining became the norm. To begin with, the residents had no steady source of income. The proceeds from chromite mining and other activities such as gold panning and contract labour on the farms were not enough to revive the lost glow the settlements had enjoyed in their 'glory days'.

Things could never have been the same, anyway. The situation of Mhangura demonstrates this reality. As Zimbizi (2001) notes, in its heyday, Mhangura Copper Mine (MCP) employed between 2,500 and 3,000 workers. After the radical downscaling, the mine finally closed in November 2000. ZMDC retrenched the remaining 1,200 workers. This adversity affected around 15,000 people who relied directly or indirectly on the mine for their livelihoods.

The situation is no different in the chromite mining areas where Save the Children (2003, 2) noted:

> Livelihoods for most households in Mutorashanga are dominated by chrome mining. Secondary food and income sources come from agricultural labouring on surrounding A1 resettled farms, wild foods and gathering, and various types of petty trade and self-employment. Since January 2003, food aid has played an important role, providing 40–50% of the food needs of the communities.

Hardly a description of a town in Zimbabwe! The incomes from alternative sources could not sustain the high level of ancillary activities and services previously provided by the mining companies. It could not sustain the level and variety of commercial services that had cropped up to take advantage of the forward and backward linkages with the then viable mining industry.

To be sure, the government did take over some of the services such as education and primary health care. But this was a government that was trying to cope with nationwide crises characterised by an economic meltdown, collapsing infrastructure and deteriorating services. When the large mining houses folded, the services almost collapsed and the opportunistic formal entrepreneurs disappeared. They were replaced by the informal sector. The process of informalisation intensified after the 'resurgence'. The settlements then became 'one large informal cesspool of everything' (Moyo 2010). They were towns, yes, but very different towns from what they had been before the decline. Also, they were very different towns from 'conventional' urban centres.

The solid infrastructure noted by Saunders (2008) fell into disrepair. The hitherto decent houses, which had been maintained by the mining companies, experienced massive overcrowding and similarly deteriorated. To be fair, the degeneration of housing, infrastructure and services was happening across the urban settlement hierarchy throughout Zimbabwe. But the mining communities had no one to hold accountable, and therefore no one willing to stop the rot. The local authorities were more interested in the communities' money than service provision (Moyo 2010). The NGOs were preoccupied with specific agendas such as emergency relief, nutrition, food aid, HIV/AIDS, democracy and other 'sexy saleable issues' (Mrs LM 2010). The embattled government had its hands full with the economic and political chaos the country had descended into, a situation largely spawned by its disastrous policies. So it is that the residents were on their own; and on their own they could not keep the settlements from crumbling into something resembling huge informal settlements; becoming a far cry from the urban areas Zimbabweans had been used to. Tellingly, a survey carried out by an urban planner showed that 73% of the 63 residents he interviewed were of the opinion that the mining settlements were not 'genuine' towns (Madziyo 2010). The most common explanation for this standpoint was that the settlements were 'not like the real towns' such as Banket and Chinhoyi.

Settlement planning and governance

The mining towns have no standard spatial pattern. This is because they have no layout plans that guide their development. There is no need for one because the local planning authorities (MRDC and ZRDC) have no planning powers on the mining compounds. According to Section 234 of the Mines and Minerals Act (Zimbabwe 1996), the mining commissioner, 'who hardly knows anything about planning' (Madziyo 2010) approves plans for certain developments on mining locations. To all intents and purposes, the mine owner is the planning authority. This has given rise to a variety of permutations in the patterns of mining settlements.

The absence of a central planning authority means there has never been any planning control except by the mining commissioner and the mining company. Unsurprisingly, national minimum housing and infrastructural standards have always not been applied in the mining towns. As is the case in most mining towns, all the land is owned by the claim owners – the mining companies. Hence, from the beginning there was no housing or land market in the mining towns. The entirety of the housing stock comprised tied accommodation. Business premises that were not operated by the mining companies were not owned by the operators, but were leased from the mining companies. In the colonial days, when 'non-whites' were not allowed to own land in 'white' areas and state land, this made no difference to the majority of the residents. However, when blacks were finally allowed to own land and property in any part of the country, and homeownership became an integral part of national housing policy, residents of mining settlements were excluded. Predictably, the issue of standards and ownership are among some of the reasons why some people believe the mining settlements are 'not like the real towns'.

The mining towns have good rail and road links to the major provincial and national urban centres. Sutton, for example, is easily accessible to Mutorashanga, the largest mining town in the northern Great Dyke. It is well connected to Banket, the largest town in ZRD. Through Banket, like all other chromite mine towns, Sutton is accessible to Chinhoyi, the provincial capital, and Harare, the national capital. The two-lane, tarred trunk road, which is still in good shape, can rival any national highway in Zimbabwe. It connects to the A1 (Harare–Chirundu) highway near the Mapinga siding, some 10 kilometres from Banket and 80 kilometres from Harare. The rail connection comes with a siding for a goods train. The good transport network was meant to make the mining locations accessible. The companies needed to be able to transport their bulky product to the market and to be able to bring in equipment and other supplies, some of which could also be bulky.

Poverty and welfare outcomes

In their heyday, the mines were zones of prosperity. In the colonial days, the standard of living was considered higher than in the commercial farming areas, communal areas, urban low-income areas and informal settlements (Mrs LM 2010). The relatively reasonable wage was complemented by free housing, electricity and water and sanitation (Sango et al. 2006). Also available were good entertainment, health and educational facilities plus a host of other services. Zimalloys' assertion that it invests in 'participating and funding various local initiatives primarily to impact positively on societal needs' (Chirasha, Shoko, and Machikicho 2006, 5) is no empty boast. As noted above, some mining companies funded successful sporting clubs, which accounts for Mhangura's legendary successes in football and athletics. With solid infrastructure, equipment, trained teachers and other facilities, the secondary schools in Mhangura, Alaska and Mutorashanga were among the best in the province. They were 'miles ahead' of the government schools originally meant for black people in the 'conventional' urban centres (Mrs LM 2010).

The good life continued for a time after independence. It was bolstered by the introduction of a statutory minimum wage by the new majority black government (but see below). Other government initiatives such as free primary education and primary health care benefited the mining communities. For a time the mining towns prospered. Then things started to change. The dramatic reversal was precipitated by the decline in world mineral prices. Retrenchments were followed by downscaling; then, in the case of chromite mining, a switch to tribute mining; and finally closures and/or abandonment. As the companies downscaled and/or closed down, and skilled workers left for greener pastures, poverty started to bite. Those hit hard were the labourers, the unskilled low-income earners. Most of them could not leave because they had known no other life apart from mining. They had no skills to offer the embattled mining industry where competition for jobs was now 'a dog eat dog affair' (Mrs LM 2010). Most of them had no rural home. So, with nowhere else to go to, they stayed at the mines.

Income from tribute mining is inadequate to cover basic needs. In addition to the exploitative contracts, 'the price paid to miners for chrome lags very far behind the rate of inflation' (Save the Children 2003, 2). This has forced tribute miners to look for secondary food and income sources. As noted above, they have diversified into 'agricultural labouring on surrounding A1 resettled farms, wild foods and gathering, and various types of petty trade and self-employment' (Save the Children 2003, 2). In the copper mining communities there was wholesale informalisation of everything: from income generation and housing to the procurement of basic necessities. In all mining settlements, gold panning and contract labour on farms have taken hold. There has been a virtual 'gold rush' to gold panning locations, especially the Angwa-Pote Basin (Kamete 2008), which is close to the copper mining settlements.

In time, the local authorities took over the administration of some of the mining settlements. Makonde Rural District Council (MRDC) took over the running of Mhangura and Alaska. ZRDC took over the running of the chromite mining settlements. However, like everything run by the public sector during this period, the situation deteriorated. The quality of housing, healthcare and water and sanitation, as well as roads and other infrastructure plummeted due to a combination of resource constraints, corruption and mismanagement. Significantly, Mhangura started experiencing chronic severe water problems as the MRDC failed to maintain service levels (Manyanhaire et al. 2009). MRDC found the task overwhelming because it was 'institutionally, technically and financially unprepared and under-equipped' (Sango et al. 2006, 195). The story has been the same in Alaska and Shackleton. Though still bad, it has been rather better on the Great Dyke where Zimalloys and ZIMASCO maintain some presence through the tribute mining arrangements (Mrs LM 2010).

The mining communities are now 'swimming in poverty' (Mrs LM 2010). In a survey of 'wealth groups', relief organisations (IPA-SC 2001, 7) found out that 55–65% of the households in Shackleton were classified as *varombo* (poor), 20–25% as *varipakati* (middle) and 15–20% as *varinani* (better-off). In Mutorashanga and its satellite towns the wealth breakdown was: *varombo* 10–20%; *varipakati* 65–80%; *varinani* 10–15% (Save the Children 2003, 9). A study by Sango et al. (2006) revealed the growing problems of poverty in Mhangura. Logically, as the economic crisis deepened and victims of the FTLP and OM/RO joined the mining communities, the percentages tilted heavily in favour of the *varombo*. With no dependable source of income from the copper mines, things have been 'leaping from bad to worse' (Moyo 2010). The chromite mining areas have not fared much better. Technical reports revealed that indicators of well-being were plummeting (Zimbizi 2001). One report concluded that 'assessments found that the area was chronically poor, and getting poorer, mainly because of mining incomes failing to keep pace with inflation' (Save the Children 2003, 9). Another report found that global acute malnutrition, severe acute malnutrition and under-five crude mortality rates in the chromite mining areas were kept within acceptable levels only because of intervention by relief organisations through food aid and primary health (Save the Children 2003). Without external intervention, the situation could have been 'calamitously worse' (Mrs LM 2010).

Government policy and the fate of mining towns in Zimbabwe

The fate of the mining towns in Zimbabwe reflects the intersection between governance, the local situation and international conditions. This section will briefly peer into government policy and actions and how this has affected mining, urbanisation and poverty. The policies include government controls, structural adjustment, ruinous economic policies and political repression. This discussion will not debate the morality of the policies, as the aim is to explain their impact on mining, urbanisation and poverty. Two of the state actions and policies, namely the FTLRP and OM/RO, have already been discussed above. As already mentioned, the FTLRP and OM/RO boosted the populations in the mining towns as each event generated an influx of IDPs into the towns.

Shortly after independence the new government imposed minimum statutory wages in all sectors of the economy. At the same time, new labour laws made it difficult to fire or lay off workers. There were also policies that made it mandatory for contract workers who had served for a stipulated period to be offered permanent employment. In the mid-1980s, as world mineral prices plummeted, employers experienced viability problems. In the face of declining profits, as Maponga and Ruzive (2002, 116) correctly argue, 'these policies exerted pressures on company profits'. By the time the tortuous negotiations to allow the companies to retrench were concluded, the damage had already been done. When finally authorised, the retrenchments hit the employees hard: full-time employees were retrenched and part-time contracts terminated. Arguably, had the rigid controls not been there in the first place, the mining companies could have adjusted incrementally and the damage to viability and loss of employment would 'not have been that monumental' (Munatsi 2010). The seeds of the collapse of the mines, and with them the towns, were sown as soon as the state imposed increasing controls on industry.

In 1991, the government adopted the World Bank/IMF-imposed Economic Structural Adjustment Programme. Among the major themes of ESAP were liberalisation and deregulation (Zimbabwe 1991). ESAP sought to 'quickly deregulate and indebt a nation seen as overprotected and inefficient' (Bond and Manyanya 2002, xv). The period after ESAP was characterised by economic downturns which the state, trade unionists and some commentators were later to blame on the programme (Bond and Manyanya 2002). Deregulation meant that the local mining industry could now hire and fire at will. Predictably, the massive loss of jobs in the mining industry that followed ESAP has been linked to the programme (Bond and Manyanya 2002). Liberalisation, another strand of ESAP, relaxed, among other things, import controls. This meant Zimbabwe's industry was no longer protected. Companies could now import metal products which were cheaper than those produced locally. The chromite industry's link to the ferrochrome industry meant that it was also hit. This multiplied viability problems and increased the spectre of mass retrenchment and mine closures.

Another economically ruinous decision was made in late 1997 when the government authorised the payment of billions of dollars in unbudgeted payments to restless war veterans. On 'Black Friday' (14 November 1997), the Zimbabwe dollar plunged from Z$14 against the US$ to Z$26 to the US$ (Bond and Manyanya 2002, xi). The Zimbabwe Stock Exchange suffered a 40% crash. Importers and investors were hit hard. The local currency has never recovered. In fact it was exacerbated by the destruction wrought on the economy in the wake of the FTLRP. Among the major victims were what remained of the copper and chromite mines. It is perhaps no coincidence that MCP folded up some four years after Black Friday.

The period after 2000 saw an increase in political violence and repression. It was mainly blamed on the ruling ZANU-PF party, which had begun losing ground in national, local and presidential elections after 2000 (HRF 2005). When the violence intensified, especially during and after elections, many victims sought refuge in the 'relatively forgotten mining towns' (Moyo 2010). This was the case following three phases of political violence: the 2000 parliamentary elections, the 2002 presidential elections and the 2008 rerun of the presidential poll (Mrs LM 2010). People from resettlement areas, communal areas, occupied farms and urban centres such as Banket, Chegutu and Karoi escaped to the mining towns. There are no figures for the number of IDPs who sought refuge in the mining towns but eyewitnesses insist that they were many (Mrs LM 2010). Some did return home after the situation normalised; many who had lost everything decided to stay.

Another ruinous government policy was the indigenisation policy, appropriately described by Saunders (2008, 74) as 'a deeply flawed approach to [black] empowerment'. The Indigenisation and Economic Empowerment Act became law in March 2008. The Act decreed that 51% of businesses should be acquired by 'eligible' (read black) Zimbabweans over a seven-year period. Under the proposed Mines and Minerals Amendment Bill,[2] the government would acquire a 51% stake in foreign-owned mining companies; 25% of the stake would be acquired at no cost. Needless to say, this flawed legislation brought memories of the suicidal land seizures. The potential damage of the Act was that 'it . . . raised both foreign investor risks and local financing costs' (Saunders 2008, 76), at a time when Zimbabwe was literally broke and needed foreign investment to take advantage of the surging world mineral prices. This, in a way, might explain why when chrome and copper prices are going up, there is no investment in the beleaguered mines. Unless the government sees reason and relents, it is difficult to imagine the mining towns 'becoming viable and springing back into serious production' (Munatsi 2010). With investors uncertain and scared of pouring money into the beleaguered chromite and copper mines, it is a given that the mining towns will remain in the quagmire of poverty, decline and degeneration.

The post-2000 economic crisis affected all productive sectors. Persistent shortages of foreign currency and skilled workforce, runaway hyperinflation and uncertainty, all of them results of ruinous policies, made mining difficult and meaningful investment in the struggling chromite and copper mines unlikely. Mobbs (2010, 1) aptly describes the devastating consequences of the economic meltdown on mineral production:

> Hyperinflation, which averaged 10,453% in 2007, compared with [an] average of 1,017% in 2006, . . . adversely affected mineral production operations. Companies that produced minerals or manufactured mineral-based commodities in Zimbabwe struggled to continue operations.

Mining towns were doomed as they bore the brunt of the fallout. Like everywhere in Zimbabwe, poverty and impoverishment set in. Because of their 'special vulnerability' (Mrs LM 2010), mining communities were hit particularly hard.

The majority were foreigners who had few or no assets and were not a politically significant constituency. So, they could be conveniently ignored. Calamity was averted only when relief organisations stepped in.

Conclusion

Mining towns in Zimbabwe started off very well. They prospered in the colonial period and the first decade after independence. Then, for a variety of reasons, they were hit hard when the mining operations they depended on first experienced viability problems, then inevitably downscaled, and finally mutated and/or closed. Some centres like Shackleton became ghost towns; others like Sutton and Mhangura mutated as tribute mining replaced the large-scale formal mining that had been undertaken by blue chip companies. But they were not dead yet. The post-2000 multiple crises, characterised initially by ruinous economic policies, and then an economic meltdown and political repression, gave them a new lease of life. There was some form of 'urban growth' as their population increased; abandoned houses were once again occupied, and some businesses managed to hang on. But this was not the phoenix rising from the ashes; it was a mutant. It has mutated in that it has been transformed into 'something else'; it has developed new physical characteristics. It can be described as a 'mutant' because it is different from others of its type – other urban centres – because of a change in its demographic, governance, spatial and economic attributes. What has emerged is different from what was there before. The orderly, hierarchical, paternalistic and autocratic settlement systems, efficiently managed and tightly controlled by the mining companies, mutated through twin processes of informalisation and impoverishment. While this is not different from the rest of the urban hierarchy where Zimbabwe's towns and cities are fading (Kamete 2006), it is distinctive in its intensity and outcomes.

Nevertheless, the chromite and copper mining towns in Mashonaland West have always been an integral part of the country's urban settlement hierarchy. Yet, for Zimbabweans used to a certain idea of 'town', the mining towns are special kinds of towns: inferior towns. Examining the origins, structure and governance of these settlements leads to the conclusion that maybe this is not an unwarranted label. Mining towns did not emerge as administrative centres, but as centres of accumulation, for the extraction of specific resources, for a limited period. They have always been different. Land and housing markets have never operated here; most of the national planning and housing controls have never applied here; nor have the minimum infrastructural standards been enforced here. Even after their takeover by local authorities it will be difficult to impose the national urban housing and infrastructural standards without undertaking a wholesale and expensive upgrading programme. This is simply impossible. What this suggests is that the mining settlements will continue to be different; different, yes, but no less urban.

Notes

1. Adding the output from tributors in the southern Great Dyke brings the tributors' total output to 570,000 mT, about 88% of the total national output of 650,000 mT.
2. The Bill lapsed following the dissolution of parliament in 2008. The government insisted it was withdrawn for 'consultations' (Karombo 2008).

Note on contributor

Amin Y. Kamete is Senior Lecturer in Urban Studies at the University of Glasgow. He moved to Glasgow from Bangor University's School of Environment, Natural Resources & Geography. He previously worked in the Department of Rural and Urban Planning at the University of Zimbabwe and the Nordic Africa Institute in Sweden. His current research interests are spatial planning and urban studies, focusing on planning theory and practice with special emphasis on cities, space and power in the context of development planning and management practice in urban Africa. His most recent publications focus on spatial planning and governance in the contested urban spaces of Zimbabwe. He can be contacted at: amini.kamete@glasgow.ac.uk.

References

Amin, A. 2008. The economic base of contemporary cities. In *A companion to the city*, ed. G. Bridge and S. Watson, 117–29. Chichester: Wiley-Blackwell.

Bailly, A., C. Jensen-Butler, and L. Leontidou. 1996. Changing cities: restructuring, marginality and policies in urban Europe. *European Urban and Regional Studies* 3: 161–76.

Bond, P., and M. Manyanya. 2002. *Zimbabwe's plunge: exhausted nationalism, neoliberalism and the search for social justice*. London: Merlin Press.

Brail, R.K., E.G. Bossard, and R.E. Klosterman, eds. 1993. *Spreadsheet models for urban and regional analysis*. New Brunswick, NJ: Center for Urban Policy Research.

Brand, L.A. 2001. Displacement for development? The impact of changing state–society relations. *World Development* 29: 961–76.

Bryce, J. 1970. *Impressions of South Africa*. Santa Barbara, CA: Greenwood Press.

Chirasha, J., N.R. Shoko, and M. Machikicho. 2005. Zimbabwe alloys: The first fifty years of operation, challenges, and opportunities. In *Southern African pyrometallurgy 2006*, ed. R.T. Jones, 5–8. Johannesburg: South African Institute of Mining and Metallurgy.

CSO (Central Statistical Office). 1993. *Census 1992: Zimbabwe national report*. Harare: CSO.

CSO (Central Statistical Office). 2002. *Census 2002: Preliminary results*. Harare: CSO.

Fewsnet. 2002. *Zimbabwe: Humanitarian situation report*. Harare: United Nations Relief and Recovery Unit.

Gono, G. 2005. *2005 Third quarter monetary policy statement*. Harare: Reserve Bank of Zimbabwe.

Hartnack, A. 2005. 'My life got lost': Farm workers and displacement in Zimbabwe. *Journal of Contemporary African Studies* 23: 173–92.

Hensman, H. 2007. *Cecil Rhodes: A study of a career*. Whitefish, MT: Kessinger Publishing.

HRF (Zimbabwe Human Rights NGO Forum). 2005. *Of stuffed ballots and empty stomachs: Reviewing Zimbabwe's 2005 parliamentary election and post-election period*. Harare: HRF

ICG (International Crisis Group). 2005. *Zimbabwe's operation murambatsvina: The tipping point*. Pretoria and Brussels: ICG.

IMF (International Monetary Fund). 2005. *Staff report for the 2005 Article IV consultation*. Washington, DC: IMF.

IPA-SC (Intercountry People's Aid and Save the Children). 2001. *Rapid livelihood security and needs assessment: Shackleton mine compound peri-urban settlement*. Harare: IPA-Save the Children.

Joseph, G., and K.J.A. Kundig. 1998. *Copper: its trade, manufacture, use and environmental status*. Materials Park, OH: ASM International.

Kamete, A.Y. 2006. More than urban local governance? Warring over Zimbabwe's fading cities. *African Renaissance* 3: 34–46.

Kamete, A.Y. 2007. Cold-hearted, negligent and spineless? Planning, planners and the (r)ejection of 'filth' in urban Zimbabwe. *International Planning Studies* 12: 153–71.

Kamete, A.Y. 2008. When livelihoods take a battering . . . : mapping the 'new gold rush' in the Angwa-Pote basin. *Transformation* 65, no. 1: 36–67.

Kamete, A.Y. 2009. In the service of tyranny: debating the role of planning in Zimbabwe's urban 'clean-up' operation. *Urban Studies* 46: 897–922.

Kamete, A.Y. 2012. Not exactly like the phoenix – but rising all the same: reconstructing displaced livelihoods in post-clean-up Harare. *Environment and Planning D: Society and Space* 30: 243–61.

Kamete, A.Y., A. Tostensen, and I. Tvedten. 2001. *From global village to urban globe: urbanisation and poverty in Africa: implications for Norwegian aid policy.* Bergen: Chr. Michelsen Institute.

Karombo, T. 2008. Controversial Zimbabwe mines bill lapses following dissolution of parliament. http://www.mineweb.com/mineweb/view/mineweb/en/page67?oid=45541&sn=Detail.

Knox, P.L., and L. McCarthy. 2005. *Urbanization: an introduction to urban geography.* Upper Saddle River, NJ: Pearson-Prentice Hall.

Lanjouw, S., G. Mortimer, and V. Bamforth. 2000. Internal displacement in Burma. *Disasters* 24: 228–39.

Leadbeater, D., ed., 2009. *Mining town crisis: globalization, labour and resistance in Sudbury.* Winnipeg: Fernwood.

Logan, J.R., and T. Swanstrom. 2005. Urban restructuring: a critical view. In *Cities and society,* ed. N. Kleniewski, 28–42. Oxford: Blackwell.

Madziyo, N. 2010. Personal communication. Harare, December 19.

Manyanhaire, I.O., J.T. Matewa, E. Svotwa, and E. Munhuwa. 2009. Water shortage related problems following the closure of Mhangura copper mine in Mashonaland West Province of Zimbabwe. *Journal of Sustainable Development in Africa* 11: 239–52.

Maponga, O., and B. Ruzive. 2002. Tribute chromite mining and environmental management on the northern Great Dyke of Zimbabwe. *Natural Resources Forum* 26: 113–26.

MBendi. 2011. Mining in Zimbabwe: overview. http://www.mbendi.comindy/ming/af/zi/-p0005.htm.

Mobbs, P.M. 2007. *The mineral industry of Zimbabwe – 2007.* Washington, DC: US Geological Survey.

Mobbs, P.M. 2010. *The mineral industry of Zimbabwe.* Washington, DC: US Geological Survey.

Molotch, H. 1976. The city as a growth machine: toward a political economy of place. *The American Journal of Sociology* 82: 309–32.

Moyo, V. 2010. Personal communication. Harare, November 19.

Mrs LM. 2010. Personal communication with social worker. Chinhoyi, December 17.

Munatsi, G. 2010. Personal communication with geologist. Harare, November 23.

Mutale, E. 2004. *The management of urban development in Zambia.* Aldershot: Ashgate.

Mwanza, D. 2010. Personal communication. Banket, December 14.

Parsons, N. 1993. *A new history of southern Africa,* 2nd ed. London: Macmillan.

Potts, D. 1995. Shall we go home? Increasing urban poverty in African cities and migration processes. *The Geographical Journal* 161: 245–64.

Potts, D. 2006. Restoring order? Operation Murambatsvina and the urban crisis in Zimbabwe. *Journal of Southern African Studies* 32: 273–91.

Sachikonye, L.M. 2003. *The situation of commercial farm workers after land reform in Zimbabwe.* Harare: Farm Community Trust of Zimbabwe.

Sango, I., P. Taru, M.N. Mudzingwa, and A.T. Kuvarega. 2006. Social and biophysical impacts of Mhangura copper mine closure. *Journal of Sustainable Development in Africa* 8: 186–204.

Saunders, R. 2008. Crisis, capital, compromise: Mining and empowerment in Zimbabwe. *African Sociological Review* 12: 67–89.

Save the Children. 2003. *Household economy assessment report al Resettlement areas and Mutorashanga informal mining communities Zvimba District, Mashonaland West, Zimbabwe.* Harare: Save the Children.

SPT (Solidarity Peace Trust). 2005. *Discarding the filth: Operation Murambatsvina.* Bulawayo: SPT.

The Herald. 2005. Government not punishing people. Harare, June 16.

Tibaijuka, A.K. 2005. *Report of the fact-finding mission to Zimbabwe to assess the scope and impact of operation murambatsvina by the UN special envoy on human settlements issues in Zimbabwe.* New York: United Nations.

UNDESA (United Nations Department of Social Affairs). 2007. *World population prospects: the 2007 revision*. Washington DC: UNDESA.

Weiss, L. 1999. Globalization and national governance: antinomy or interdependence? *Review of International Studies* 25: 59–88.

World Gazetteer. 2011. Zimbabwe: largest cities and towns and statistics of their population. http://www.worldgazetteer.com/wg.php?x=1185906708&men=gcis&lng=en&des=game lan&dat=200&srt=epnn&col=aohdqcfbeimg&geo=247 (Accessed 11 January 2011).

ZIMASCO. 2010. Mining operations. http://www.zimasco.co.zw/features.html.

Zimbabwe, Government of. 1991. *Zimbabwe: Framework for economic reform (1991–95)*. Harare: Government Printer.

Zimbabwe, Government of. 1996. *Mines and Minerals Act (Chapter 21:05)*. Harare: Government Printer.

Zimbabwe, Government of. 2004. *Zimbabwe: Towards sustained economic growth – macro-economic policy framework for 2005–2006*. Harare: Government Printer.

Zimbizi, G. 2001. *Study on socio-economic status of children and women on commercial farms, mining and peri-urban areas*. Harare: UNICEF.

Botswana's mining path to urbanisation and poverty alleviation

Thando D. Gwebu

Relative to the rest of sub-Saharan Africa, Botswana has recorded exceptionally high rates and levels of urbanisation. At national independence in the 1960s, Botswana was overwhelmingly a rural country. Now, over half of the population currently lives in urban settlements. The discovery and exploitation of the country's rich mineral resources since independence in 1966 have made a significant contribution to the development of new towns, such as Orapa, Jwaneng and Selebi Phikwe. This contribution documents the nature of diamond and copper-nickel mining activities in Botswana and their influence on the urban settlement pattern. Botswana's mining towns are run on welfare capitalist lines by foreign companies, primarily DeBeers, in which the Botswana government has a minority share. The populations of the towns and surrounding areas have grown well beyond their planned levels. The government has been acutely conscious of the limited life of the mines and of the need to make provision for employment after the end of mining. The Selebi Phikwe nickel mine was expected to cease production in 2010 and efforts have been made to create non-mining employment with some success. The dilemmas posed by depleting mineral supplies and the government's attempts to find economic employment alternatives to sustain existing urban settlements are discussed. Given Botswana's reputation for investing mining wealth in infrastructural development and public welfare for current and future generations, this analysis assesses the challenges large-scale mining poses for achievement of sustainable development, social justice and poverty alleviation.

Before gaining national independence in 1966, Botswana was known as the British Protectorate of Bechuanaland. Famous for its vast Kalahari Desert, the country is located on Southern Africa's central plateau. Until the 1970s, its population consisted primarily of rural pastoralists, as well as San hunters and gatherers occupying the drier parts of the desert. With a population of about 2 million, Botswana is estimated to be growing at about 2% per annum.

Botswana is one of Africa's most urbanised countries at present. Yet at national independence in 1966, this was far from being the case. Its small, highly dispersed population became increasingly concentrated in urban settlements over the last three decades. At the same time, Botswana has gained the reputation of being the 'Norway of Africa', averting the pitfalls of the resource curse (Bebbington et al. 2008), and serving as a model example of public planning and investment in mining for the needs of its increasingly urban citizenry.

Mining has been the backbone of the country's economy since the 1980s. Several studies have been conducted on the nature and significance of the mining sector to the country's welfare but none mention its impact on urbanisation. At an early stage, Colclough and McCarthy (1980) addressed three issues related to how the Botswana government could promote mineral development in the country: first, how the supportive infrastructure for the mines could benefit other economic sectors; second, how the fiscal arrangements for the mines could maximise financial flows to government revenues; and third, ways to minimise regional imbalances resulting from mineral development.

A decade later, Harvey and Lewis (1990) focused on mineral policy and mining development in terms of: issues internal to the mineral sector; major links of mining to the rest of the economy; mineral economics; the scale of mineral development and its importance in the aggregate economy; the evolution of mineral policy and legislation; and the role played by mineral negotiations. Gaolathe (1997) examined the growth of the mineral sector in Botswana and the extent to which the sector could be made more beneficial to the continued socio-economic development of the country.

It is only in the past 10 years that the literature has addressed the interaction of mining and urbanisation. Gwebu (2006) investigated the role of mining settlements in the context of a differential urbanisation model and Botswana's unique environmental, demographic and socio-economic conditions between 1981 and 1991. Most recently, Ntsabane et al. (2010) have examined the economic and environmental impacts of the Jwaneng diamond mine on the residents of the town and its surroundings, showing that while the mine has generated a new township, infrastructure and town-bound migration, it has failed to induce development in the smaller villages of the Southern District.

This article goes beyond the story of Botswana's enormous windfall gain of diamond discovery to document how mining has catalysed urban growth and how urbanisation has been channelled to achieve poverty alleviation, while taking account of depleting mineral supplies over the long term. The next section will review the country's mining history, followed by an outline of Botswana's urban development trajectory, before moving to an in-depth analysis of the urban settlement hierarchy and differentiated urban growth patterns. Thereafter, the role of the state in managing mining and urbanisation will be examined, followed by consideration of the significance of the employment impasse associated with mining, before concluding.

Mining history

Predating the discovery of diamonds, modern mining towns in Botswana first surfaced over a century ago when gold was discovered in Francistown. During the 1950s and early 1960s, mining activities encompassed gold, manganese and asbestos production. However, the scale of operations was small relative to the present. By and large, today's mining industry reflects the Botswana government's efforts to diversify the country's economy after the severe drought and consequent cattle deaths of the early 1960s that crippled beef production, hitherto the backbone of the national economy.

Botswana's human settlement landscape (Figure 1) encompasses: first, the central places of immigration of what can best be referred to as the 'core regions'; second, upward transitional areas and resource frontiers such as diamond mining towns; third, downward transitional zones such as rural areas and the ill-fated Selebi Phikwe copper-nickel mining town; and, fourth, the development corridors that link the various rapidly growing segments of the space economy. The core, upward transitional areas and development corridors are situated in the eastern segment of the country that is ecologically best endowed with good rainfall, soils, well-developed physical infrastructure and socio-economic facilities.

Botswana's mining history has been episodic. Figure 2 shows the location of the country's mining towns. In 1866 gold was discovered in the north-east of Bechuanaland, where mining commenced in 1869 in an area occupied by the BaKalanga.

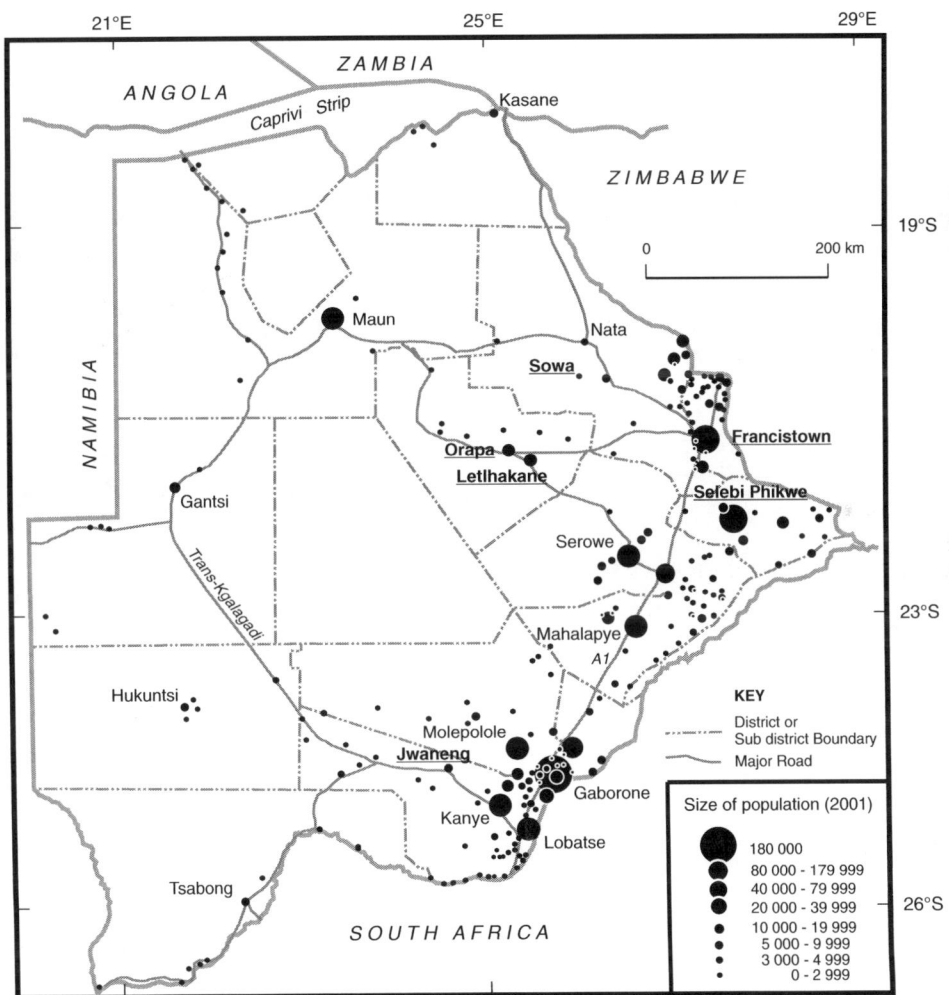

Figure 1. Human settlements and population density.
Source: Department of Environmental Science, University of Botswana 2009.

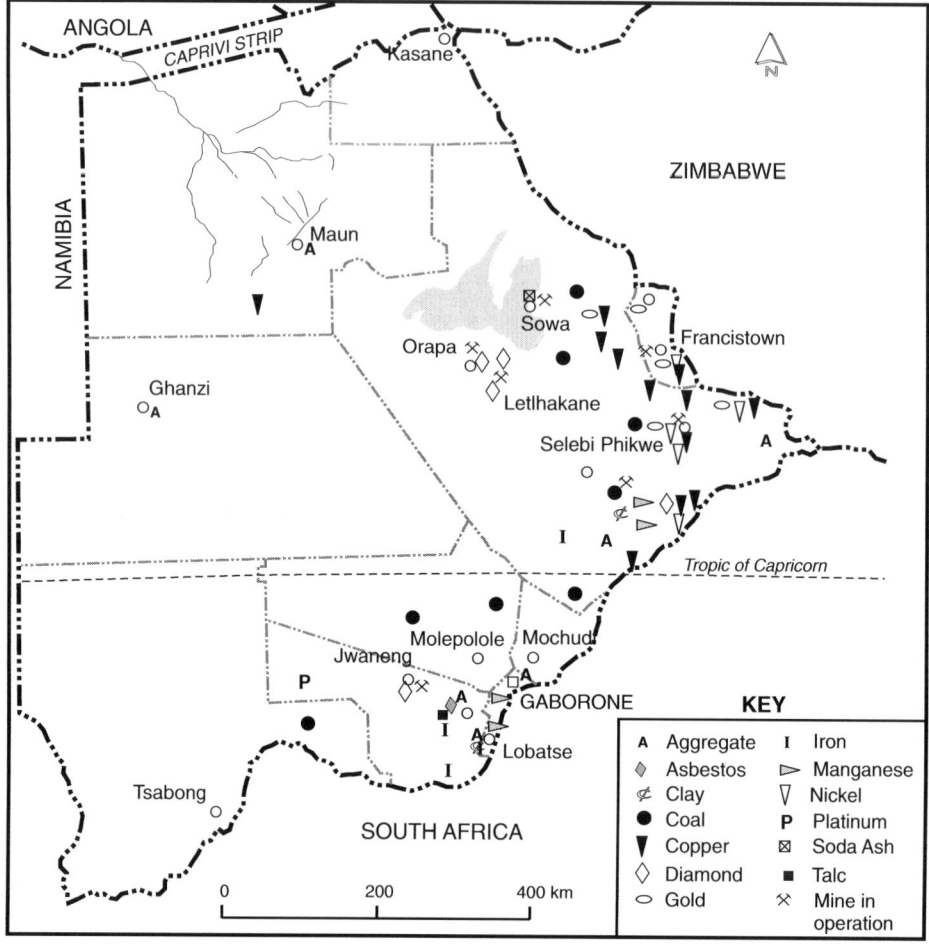

Figure 2. Minerals and mining towns.
Source: Department of Environmental Science, University of Botswana 2009.

A few years later, King Lobengula of the Ndebele, who had displaced the BaKalanga, granted the Tati Concession for all minerals extending over a 5000 km^2 area (Colclough and McCarthy 1980).

The Tati gold rush attracted miners from South Africa, Europe and Australia, in addition to migrants from the mine's hinterland. Once it had become apparent that the quantity of gold reserves had been wildly exaggerated, initial euphoria dissipated. Foreign migrants were faced with three choices: return to their countries of origin, relocate to South Africa or explore alternative locally available means of livelihood. Most opted for the last course of action (Gwebu 1987). The influx of white fortune-hunting prospectors over the period from 1880 to 1945 gave rise to Francistown, a small urbanised settlement associated with the Monarch mine site.

In 1959, the mine's output had declined to 6.2 kilograms of gold but it continued to produce in a desultory manner until 1964 when the Monarch mine was closed (Silitshena and McLeod 1989). In this context, the British Economic Survey mission made its pre-independence report on future prospects for economic development in Botswana, observing that there was no mining of note in the territory, nor were there optimistic prospects of mineral deposits for future exploitation (Colclough and McCarthy 1980).

Post-independence geological surveys, however, identified the existence of several minerals, including gold, silver, lead, zinc, antimony, tungsten, chromium, iron, soda ash manganese, copper, nickel, coal, diamonds, copper-nickel, coal, soda ash and salt. The emphasis of this contribution will be on the two leading mineral sectors, namely diamonds and, to lesser extent, copper-nickel.

The discovery of diamonds by De Beers at Orapa in 1967 marked a turning point. The diamond fields, extending over 112 hectares, proved to be one of the largest known diamond-bearing kimberlite pipes in the world. Further exploration led to the discovery of smaller diamond pipes at Letlhakane mine, situated about 30 km south-east of Orapa. Production at Orapa, under the Debswana Diamond Company, began in 1977. In 1975, other kimberlite pipes of larger size and superior quality were discovered at Jwaneng, 125 km west of Gaborone, which began production in 1982. The youngest is the Damtshaa mine, which became operational in December 2002.

The 1970s were mineral boom years, which kick-started the transformation of the Botswana economy. At national independence in the mid-1960s, agriculture was the main sector, contributing over 40% of GDP while mining contributed only 2%. By 1985 mining's contribution was 47% and foreign earnings were overwhelmingly mineral-based. Diamonds' contribution to external trade earnings reached 80% in the early 1990s and 88% in 2001.

Currently Botswana produces diamonds, copper-nickel, coal, soda ash and salt. Diamonds have been the basis of export revenue since their discovery in Orapa. Most of the mining is open cast but this will change to underground mining once a depth of 35 metres is reached. Open cast copper mining started in 1974 at Selebi Phikwe but has now switched to pit mining at deeper levels. The main coal mine is at Morupule near Palapye. It was built to provide coal for the power plant at Selebi Phikwe's copper-nickel mine. Coal mining takes place underground through highly mechanised processes. Production of soda ash and salt started in 1991 near Sowa Pan. The soda ash and salt occur as brine under the pan surface. The extracted liquid is pumped into evaporation ponds to crystallise before harvesting.

Migration to mining settlements, particularly Selebi Phikwe and Jwaneng, has been a central impetus to Botswana's urbanisation. Individuals have moved from rural hinterlands to mining centres for formal and informal employment. Open mining towns have seen the spontaneous development of squatter settlements arising from an inadequate housing stock and services for the migrants. By contrast, other mining towns, such as Orapa, have an enclave nature; closed for security reasons to prevent diamond smuggling.

Modern mining activities are a joint effort of government, multinational corporations and donor agencies. Soon after independence, the state vested mineral rights in the national government in order to ensure that benefits from mineral development would accrue to the whole nation and to ascertain effective control over mining activities. The Mineral Rights in Tribal Territories Act (1967) vested mineral rights in Communal Lands to the national government whereas The Mines and Mineral Act (1967) governed prospecting and mining development rights. The latter Act was revised in 1976 such that holders of prospecting licences had to carry out agreed work programmes and annual expenditures. After a maximum of seven years, a prospecting company had either to re-apply for a mining lease or surrender the rights to the area. The Precious and Semi-Precious Stones Protection Act (1968) provides for the protection of the precious stones industry and regulation of dealings in precious stones and semi-precious stones.

Mining corporations are expected to make welfare provisions for their employees and hence mines have a distributive effect on national welfare through public spending that is directed at the provision of services and infrastructure mainly for the Batswana people. Many mines provide healthcare to their surrounding rural areas. Mine doctors and nurses visit nearby village clinics regularly and seriously ill rural patients receive specialised treatment at mine hospitals. On the other hand, some communicable diseases, notably AIDS and tuberculosis, are associated with mining.

Botswana has a limited small-scale mining industry and needs to introduce structures to support the sector (Matshediso 2005). The Mineral Policy does not address issues relevant to the small-scale mining sector.[1] Batswana have frequently criticised government for not facilitating their entry into the multi-billion pula mining industry. Citizen entrepreneurial participation has been limited to holding prospecting licences and earning limited equity in mining projects through joint venture arrangements.

The capital- and technology-intensive nature of large-scale mining in Botswana limits links with the rest of the economy. Forward linkages process and fabricate output, adding value and creating local employment, but only a few diamond polishing ventures have been established. Final demand linkages relate to domestic spending of profits and wages generated by the mines. Fiscal linkages involve taxation and revenues from the sector and how they are used. Only the fiscal linkage has generated substantive income from mining, which will be discussed later.

Mineral deposits have usually been discovered in remote areas used for cattle posts and arable fields, as was the case in Jwaneng, Letlhakane and Selebi Phikwe. After discovery, existing local residents are requested to relocate in order to pave the way for mining operations. Although compensated for lost assets and relocation expenses, this disrupts the basic rights and freedom as well as the social and cultural life of residents. For example, the relocation of the Basarwa people from the Central Kgalagadi Game Reserve in 1983 for diamond exploration compelled them to switch from hunting and gathering to pastoral and arable farming. Those who remain in the new settlements have become demoralised, with many facing idleness, alcoholism, despair and heavy dependence on government handouts (Taylor and Mokhawa 2003).

Two ore bodies containing copper, nickel and cobalt were discovered at Selebi Phikwe in 1966. The mine started full-scale production in 1974 as a joint venture between a holding company for copper-nickel mining, the Bamangwato Concession Limited (BCL), jointly owned by the Botswana government (with 33% shares) and Anglo-American Corporation and American Metal Climax (AMAX). BCL faced many technical and financial problems. For instance, the compensation that was to be paid to the affected people was delayed until 1973 and the people refused to move out of the area, a date that coincided with the construction of the mine infrastructure and the town (Asare and Darkoh 2001). These unhappy circumstances exerted a push factor on rural people to migrate to the urban areas.

Urbanisation

Botswana's main national core region, Gaborone, constitutes the hub of markets, large-scale commercial enterprises, industry and innovation. Francistown, the major regional centre in the north, and the modern 'urban villages' of Serowe, Palapye, Mahalapye, Mochudi, Molepolole, Ramotswa and Kanye are characterised by increasing investment in urban activities. They were former large 'agrotowns' that have now acquired the structural and functional attributes of urban areas.

In Botswana, a minimum population threshold of 5000 residents defines an urban settlement, with at least 75% of the economically active population engaged in non-agricultural activities. Table 1 depicts urbanisation levels from 1971 to 2001, 1971 representing the pre-mining boom era. The urban growth rate leapt to 12% per annum between 1971 and 1981, doubling the level of urbanisation, then rose even higher to 14% in the next decade, pushing the urbanisation level to 45% before the growth rate levelled off at 4% between 1991 and 2001 (Republic of Botswana 2001).

Mining was a major stimulus to urbanisation. Urban growth was taking place in four main ways. First, as can be observed from Table 1, the number of places categorised as urban increased. Second, and most significantly, urbanisation resulted from a tidal wave of rural-to-urban migration, in response to income disparities between the urban industrial sector and the rural agrarian sectors, as the impact of mineral production pumped through the national economy. Low agricultural commodity prices, weak agricultural marketing systems and policies, periodic droughts and desertification propelled the rural population towards urban centres just as private and public sector investment connected with mining was expanding in the urban areas. Third, termination of the formal international migratory labour system of Batswana miners to South African mines added to the urban migration flow.[2] Fourth, the economic booms of the 1980s and 1990s were accompanied by increased urban employment opportunities, especially in the construction and service sectors, in connection with a relatively strong and stable currency and the localisation and reservation of jobs for Batswana nationals.

Table 2 shows that there were only five modern towns in Botswana in 1971. Of these, three had their origins associated with mining activities. Francistown was the earliest, having been established in the late nineteenth century, as described in the preceding section. As its mineral wealth declined its economy gradually transformed from mining to transportation, commerce and light industry.

Between the 1966 discovery of copper ore and 1981, the population of Selebi-Phikwe increased six-fold while that of Orapa more than quadrupled. The latest

Table 1. Urbanisation levels, 1971–2001.

Year	No. of urban places	Total urban populations	Total national population	Urban as a % of total populations
1971	5	54,300	596,900	9.1
1981	8	166,400	941,000	17.7
1991	25	600,100	1,326,800	45.2
2001	34	909,800	1,680,900	54.1

Source: Central Statistics Office 2001 (Republic of Botswana 2001).

Table 2. Growth of population in urban settlements, 1971–2001 ('000).

Town	1971	Growth rate (%)	1981	Growth rate (%)	1991	Growth rate (%)	2001
Gaborone	17.7	12.9	59.7	8.4	133.5	3.4	186. 0
Francistown	18.6	5.3	31.1	7.7	65.2	2.4	83.0
Lobatse	11.9	4.8	19.0	3.2	26.0	1.3	29.7
Selebi-Phikwe*	4.9	19.7	29.5	3.0	39.8	2.3	49.8
Orapa*	1.2	15.8	5.2	5.4	8.8	0.4	9.2
Jwaneng*			5.6	7.2	11.2	3.1	15.2
Letlhakane*					8.6	5.7	15.0
Sowa*					2.2	2.8	2.9
Total	54.3	10.7	150.1	7.0	295.3	2.8	390.8

*Mining towns.
Source: Central Statistics Office 2001 (Republic of Botswana 2001).

entrant to the system was the mining town of Jwaneng, where diamond kimberlite pipes were discovered in 1972. Thereafter, between 1981 and 1991, the number of modern towns increased from six to eight, of which the latest entrants were the diamond mining towns of Letlhakane and Sowa where soda ash and common salt were discovered. The growth rates in Table 2 show a pattern in which settlements with newly discovered wealth grow very rapidly. However, over time their rates slow down to below the rates prevailing in Gaborone and Francistown.

Urban settlement hierarchy and its differentiated growth pattern

Gaborone, the national capital, far from the mining centres of the north, nonetheless has grown extremely rapidly following its founding at independence in the 1960s when the nation's administrative capital was moved from Mafeking in South Africa to just across the border in Botswana. Interestingly, the national capital has retained its urban primacy throughout the mining boom period (Table 3).

Gaberone experienced an especially high growth rate (12%) during its first decade of existence because it was the focus of national public and private investment and employment opportunities. Its growth has declined over the years due to the suburbanisation of its dormitory satellite communities, coupled with the attraction for investment and the emerging importance of secondary and tertiary centres in the settlement hierarchy. Table 3 provides a listing of the major settlements of Botswana comparing their growth decade by decade.

Mining boom, 1971–1981

After the discovery of diamonds and investment in their exploitation, the mining towns, along with Gaborone and its satellite centres, recorded the highest growth rates respectively. The population growth in Gaborone's suburbs and satellite centres grew at 8% per annum during that decade relative to the mining towns with the astronomical growth rate of 21%. Figure 3 graphically depicts the growth rates experienced by the respective centres.

Table 3. Settlement hierarchy, population totals and growth rates

Description	Pop. 1971	Growth rate	Pop. 1981	Growth rate	Pop. 1991	Growth rate	Pop. 2001
Mining towns[1]	6,100	18.9	40,300	4.3	62,000	2.2	77,100
Primate core – Gaborone	17,700	12.2	59,700	8.1	133,500	3.3	186,000
Subcores within 10 km radius of primate core[2]	5,253	6.3	9,825	10.0	26,700	7.7	57,400
Satellite dormitory suburbs within 75km radius of primate core[3]	56,912	6.5	108,841	4.6	171,997	3.0	232,600
Peripheral major regional centres over 200km radius from primate core[4]	76,610	4.9	125,111	5.4	215,000	3.6	308,700
Rural service centres[5]	12,425	5.3	21,178	4.3	32,397	3.9	48,000

Notes:
1. Selebi Phikwe, Orapa, Jwaneng, Sowa.
2. Tlokweng, Mogoditshane.
3. Mochudi, Molepolole, Ramotswa, Lobatse, Gabane, Kanye, Moshupa, Thamaga, Kopong.
4. Francistown, Palapye, Serowe, Mahalapye, Maun, Letlhakane, Kasane, Ghanzi, Bobonong, Tonota, Tutume.
5. Letlhakeng, Lerala, Shoshong, Mmadinare, Maitengwe, Gumare, Tsabong.
Source: derived from the Central Statistics Office data.

Jobs were expanding in this period, triggering migration from the countryside. Being the first decade of discovery and exploitation of diamond resources in Botswana, the mining centres attracted a lot of population away from the rural and peripheral regional hinterlands.

Secondary growth phase, 1981–1991

In this phase, growth in and around the primate city took precedence. Growth was especially high in the surrounding sub-core, which benefited from the commercial and industrial spillover effects and the primate city's agglomeration diseconomies in the form of traffic congestion and shortages of land and housing. Exclusive residential areas such as Mokolodi and Phakalane sprang up just outside the capital city.

Meanwhile, mining town growth slowed down because of retrenchments related to the impending closure of the major copper mine at Selebi Phikwe. Many of the retrenched made their way to Gaborone. Restricted access to some diamond mines and the inherently low absorptive capacity of the mining industry, because of its high capital–labour ratios, contributed to the slow growth.

The peripheral sub-regional centres recorded growth rates that were second only to those experienced by the sub-cores. This could have been due to a large village upgrading programme implemented in the newly reclassified urban villages, coupled with the decentralisation of investment and services to local district authorities. The upgrading of infrastructure attracted government and private sector enterprises to the urban villages and created many employment opportunities in the district-level capitals. This same effect was created with the establishment of sub-regional centres at rural outposts.

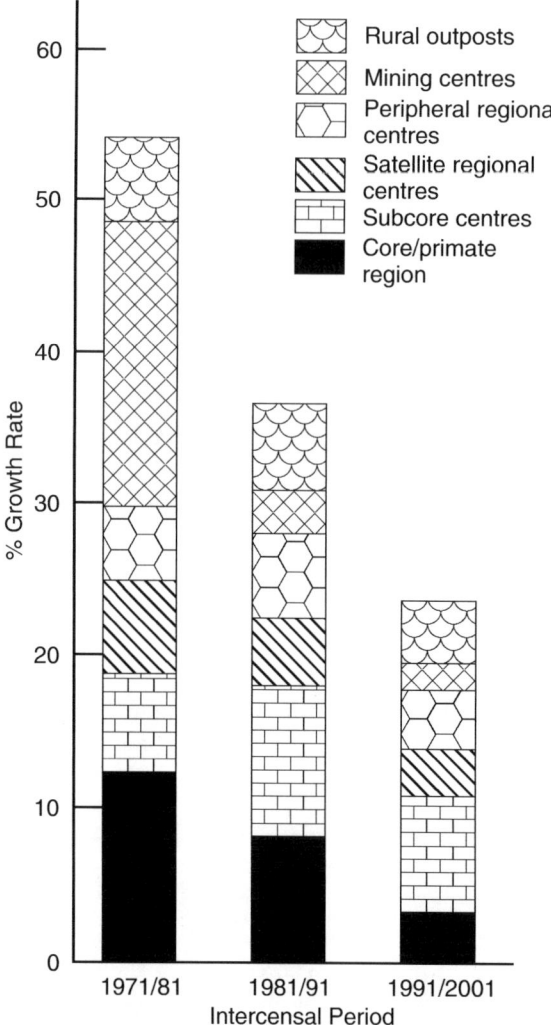

Figure 3. Cumulative annual growth of settlement types over time.
Source: derived from Central Statistics Office data.

The sub-cores' attraction was mainly due to sub-stream migration from Gaborone's settlement core, coupled with migration from the satellite, rural and mining centres. Sub-regional peripheral centres also expanded quickly due to stepwise migration in response to decentralised economic opportunities. These patterns characterise the advanced primate and early intermediate city stages in the evolution of Botswana's urbanisation process.

The peripheral centres represent strong polarisation reversal as they replicate the industrial, commercial and administrative role of the core. Their growth is also likely to have been accelerated by severe drought and environmental refugees who were forced to abandon small isolated settlements and move to larger settlements where they could obtain food rations, work, reliable water and dependable social services.

Third wave growth of the periphery, 1991–2001

Between 1991 and 2001, the settlements that experienced the highest intercensal growth rates were sub-core centres, rural outposts and peripheral regional centres. The relative growth rate of Gaborone's core continued to shrink whereas sub-core growth remained comparatively robust because of industrial and commercial enterprises' relocation from the core to the sub-core and from satellite communities and elsewhere within the settlement hierarchy.

Mining towns recorded the least growth, reflecting the slump in demand for diamonds closely linked to the world economic recessions of the 1980s and 1990s and the effects of the ongoing closure of copper-mining at Selebi Phikwe. Figure 4 compares the quite marked differences in spatial growth between the three decades illustrating Perroux's (1970, 94) observations that: 'growth does not appear every-where at the same time; it becomes manifest at points or poles of growth, with variable intensity; it spreads through different channels, with variable terminal effects within the whole economy'. Population concentration started with the primate city, Gaborone, and subsequently spread outwards with a ripple effect, first to the nearby intermediate centres and then to those farthest out, before reaching the smallest ones (Gwebu 2006).

Agglomeration and other scale economies sustained cumulative multiplier causation effects that concentrated capital, labour and other factors of production in the capital city. Zelinsky (1971) has referred to this phase of mobility as the early

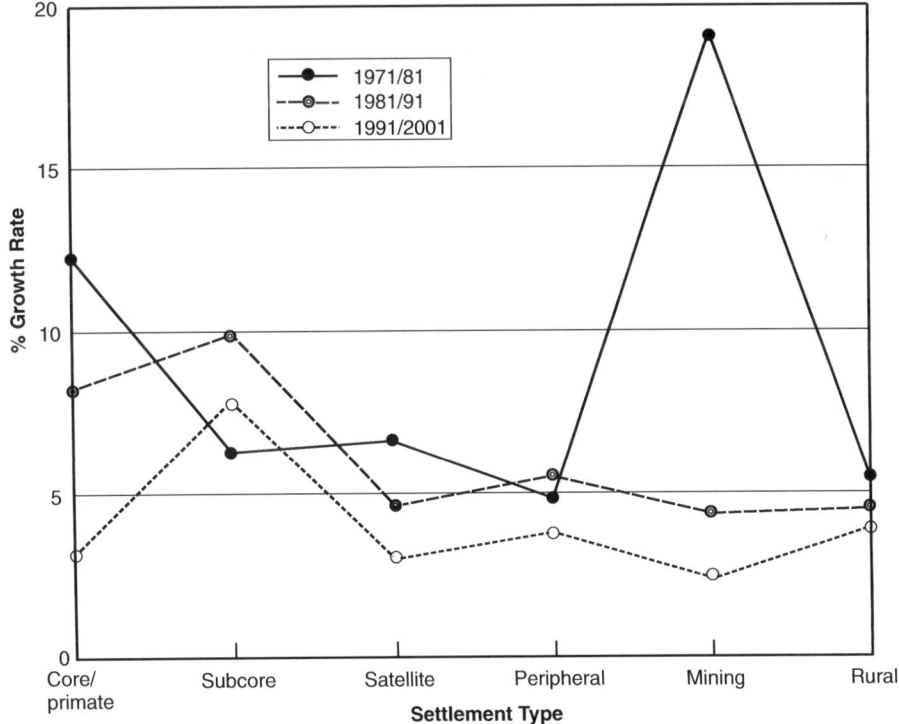

Figure 4. Comparative settlement annual growth rates over time.
Source: derived from Central Statistics Office data.

transition stage, whereas Kontuly and Geyer (2003) call it the concentration phase. The saturation limit of this phase has been reached and is clearly defined by symptoms of agglomeration diseconomies and over-urbanisation. These have led to centrifugal sub-stream migration from the primate city, in search of lower input costs, cheaper land, accommodation and the environmental amenity offered by the intermediate and nearby centres. Government-led investment in physical infrastructure and social services has also attracted both the private sector and population from the lower-order centres to these intermediate centres.

Demographic growth of these centres and that of the capital has generated concomitant development and spatial fission-cum-sprawl most visible along the major arterial radial routes and then in interstitial areas. This growth pattern of intermediate centres, near the capital city, has been replicated in those farther afield. Such growth has marked polarisation reversal. Data from the latest census suggest that the smallest resource frontier settlements have experienced mercurial growth over the past decade. Such growth involves skilled personnel moving from the higher-order centres, complemented by the relocation of unskilled and semi-skilled migrants from their hinterlands.

The annual growth rate of mining centres peaked between 1971 and 1981, due to mainstream migration from the rest of the space economy: ever since, mining towns have been characterised by declining growth associated with periodic global recessions. Migrants' destinations from this sector have been the sub-core centres, peripheral regional centres and the rural outposts.

The growth of the lowest-order centres of the urban hierarchy reflects the spread effects arising from political decision making and de-agglomeration operating at higher levels of the urban system. Figure 4 dramatically illustrates the mining town boom during the 1970s and how growth in Gaborone core has given way successively over the decades to the sub-core.

The role of the state in managing mining and urban growth

Planning

Relative to other countries in Africa, Botswana's mining and urbanisation has been subject to planning structured by defined development objectives. At independence, the country was among the poorest nations in the world and episodes of drought were decimating the pastoralists' cattle production, which constituted its economic base. Standards of living for the country's citizenry have markedly improved since then. Botswana is a participatory democracy subject to prudent management of the macro-economy. The evolution of the mining sector in Botswana reflects the government's efforts to play a central role in shaping industrial policy measures.

Through its successive national development plans (NDPs), the government has stipulated specific policies regarding: the promotion or acceleration of mineral prospecting and new mine development; maximisation of economic and financial benefits resulting from mining operations; encouragement of activities that generate real value-added and linkages with the rest of the economy; diversification of the economy; creation of employment and training opportunities for citizens; and minimisation of environmental damage by mining operations (Gaolathe 1997).

State revenue and expenditure

Mining provides taxes, dividends and royalties to the Botswana state. The impact of mining on the economy includes net foreign exchange inflows, provision of technical skills and absorption of domestic resources. The construction of dual purpose infrastructure such as communications, power and health facilities has stimulated growth in other sectors of the economy as well as facilitating the movement of goods and services and the population, and the diffusion of innovations to other nearby places.

The state has invested returns from mining in rural development. For example, through the Accelerated Rural Development Programme, more than 20 million pula was spent on the construction of rural infrastructure, marketing, labour-intensive manufacturing, credit extension, primary schools, clinics, roads and water supplies in the early 1970s. Rural expenditures increased four-fold between 1973–1974 and 1974–1975 and this higher rate of rural investment was maintained. In 1996–1997, government obtained 48% of its revenue from mining and derived taxes worth 3.3 billion pula from that sector (Silitshena and McLeod 1998). The money was used to finance agricultural extension and marketing through the Arable Lands Development Programme (ALDEP) and services such as housing, health and education, post offices, police stations, banks and customs facilities.

Long-term sustainability and diversification

A report by the IMF into the long-term sustainability of Botswana's reliance on diamond mining for its fiscal stability pointed out that, at present rates of mining, the country's diamond resources could run out by 2029 (Basdevant 2008). This sets serious long-term challenges that the government is now confronting by continuing prospecting for minerals and establishing new mines. Most importantly, the government is using revenue from mining to diversify the economy.

Initially, mining was a form of economic diversification from an economy heavily dependent on exports of beef and migrant labour remittances. Subsequently, government revenue, especially from diamond mining, has been used to build infrastructure and to provide education and training. In this way, the natural advantage of mineral resources has been converted into the basis for more diversified economic growth (Republic of Botswana 2009, 27). At the macro-economic level, the government's expenditure strategy has aimed at preventing the non-mining sectors of the economy becoming uncompetitive and unprofitable. This has enabled the economy to make significant progress in economic diversification. One indicator of diversification success is that Botswana has become significantly less dependent on imports. As a small economy with a small population, imports remain vital. Nevertheless, the ratio of imports of goods and services to GDP has fallen very substantially. In the 1970s, it was just under 70%. By the first years of National Development Plan 9 (NDP9) it sank to under 38%, indicating remarkable structural change (Republic of Botswana 2003, 28).

Economic diversification has also been directed at those mining ventures that appear to be faced with closure. Selebi Phikwe copper-nickel mine is a classic example, as outlined below.

Selebi Phikwe's economic diversification trajectory

Selebi Phikwe mine was conceived as part of a major economic diversification programme known as the Shashe Project aimed at expanding the national economy away from the traditional economic activities of pastoral farming and subsistence cultivation (Asare and Darkoh 2001). It involved the construction of Shashe dam, over 80 km of pipeline to Selebi Phikwe mine, a 60MW power station and a railway line linking the mine to the Serule main rail route.

The town grew to become an important employment, administrative and commercial centre for the surrounding area in the Central district. The government took responsibility for designing and servicing the township because it sought to create an integrated local community, in contrast to the closed diamond town of Orapa.

The hitch has been that the prices of copper-nickel have fluctuated a great deal since the 1970s, causing the mine to operate at a loss. This was exacerbated by labour unrest that culminated in a major strike in 1975. Copper prices rose in 1988 only to fall sharply in the early 1990s. Other problems plaguing Selebi Phikwe mine included its relatively remote location, high transport costs, its distance from the seat of government in Gaborone and lack of financial and commercial–industrial services (Silitshena and McLeod 1998). Selebi Phikwe was in danger of becoming a ghost town due to the unprofitable operation of the mine and the company's accumulation of financial debt.

Cognisant of its social responsibility, the Botswana state intervened to ensure that the mine remained open. Since 1988, it has developed an industrial site to encourage businesses to set up light manufacturing industries and established the Selebi Phikwe Regional Development Programme (SPRDP), in order to promote non-mining economic activities in the town and surrounding areas. A government-sponsored special support package offered financial support to assist firms with costs for labour, training and marketing for the first five years of investment, a 15% reduction of corporate tax for the first 20 years of operation, cheap land, establishment of a housing development programme and strategic road building.

The Selebi Phikwe Regional Development Promotion Unit helped draw up a special incentive package (SIP) for firms located in the town which had been in existence for at least 10 years and were willing to invest at least 25% of the project's fixed and working capital. Of the 17 companies that had located in Selebi Phikwe by 1995, only one managed to meet these criteria and obtain SIP incentives. However, since 1994, locational incentives such as concessional taxes and blocks of land have become universal in Botswana, thereby devaluing Selebi Phikwe's incentive package at the same time as the component of the World Bank loan that was used to fund the unit was drawn down.

The African Growth and Opportunity Act (AGOA) of 2000 was intended to prime the development of the apparel industries. The Act was meant to permit the export of manufactured textile products from Botswana to the United States, with minimal quota limitations. The Selebi Phikwe Regional Development Programme attracted some industries and some businesses such as textile manufacturing, bulk plastics, wholesaling, retailing, restaurants and hotels. Selebi Phikwe mine has played a leading role in employment generation, particularly within its immediate environs.

The positive impact in employment generation has contributed to the improvement of the living standards of the local people (Asare 1999).

The largest number of jobs was created in the textile and apparel sector as a result of AGOA. In 2002, AGOA exports from Botswana increased almost four-fold over the previous year, driven by an increase in apparel exports. According to officials of the Selebi Phikwe Trade and Investment Promotion, employment in the manufacturing sector increased from 139 people in 1981 to 6975 in 1998 (Asare and Darkoh 2001), which constituted a success story in the government's efforts to create non-mining employment opportunities as well as to diversify the economy by boosting the manufacturing sector. The town was chosen as a priority centre for regional industrial development and is a principal location for large-scale light manufacturing.

Various amenities complement its productive infrastructure. Selebi Phikwe has tarmac roads leading to the town and within the town there are libraries, parks, petrol stations, shopping malls, a golf course, sports and hobby clubs, service organisations, educational institutions, hotels and growing industrial complexes. An airport is being upgraded to cater for the equivalent of Boeing 737 aircraft. Good roads connect Selebi Phikwe with major centres to the north and south. Selebi Phikwe and its surroundings boast 42 tourism facilities, including hotels and guesthouses and 26 lodges and camps.

The available health facilities are the district council hospital and clinics or health centres (Asare and Darkoh 2001). These facilities are provided by the government but managed by the Town Council. BCL Mine has its own hospital but access is restricted to its employees and their close relatives. Socio-economic benefits such as the development of social and economic infrastructure, manufacturing and construction industries, commercial and public sector activities have improved significantly as the government's commitment to diversify the local economy is high.

Selebi Phikwe has attracted skilled and unskilled labour from the hinterland and other parts of Botswana. This relates in part to its urban markets and the purchasing of food supplies from their service areas (Colclough and McCarthy 1980). Employees of the informal sector who migrated from the nearby villages have benefited from mining by selling poultry, beef, vegetables, fresh milk, traditional beer, firewood and craft items to the mineworkers. Commercial activities boomed for the local citizens who were involved as small-scale unregistered traders, tailors and hairdressers. Some of the migrants became involved in construction, vehicle repairs, welding and fabrication. However, to date most of the large-scale investors, for whom the Botswana Development Corporation had constructed factory shells, regard Selebi Phikwe as remote and have preferred to locate in the more accessible and densely populated areas in the south-east of the country.

Mining and national welfare

At independence in 1966, Botswana was one of the poorest countries in the world, by almost any indicator. GDP per capita was less than US$100 (Figure 5). Driven mainly by the expenditure of minerals revenue, but also by private sector investment in a relatively favourable investment climate, the Botswana economy grew extremely fast. In the 42 years to 2006, the real growth of GDP averaged 8.8% annually (NDP10, 27). Today the country has achieved the status of a middle-income economy with a GDP per capita purchasing power parity of $15,800, which is

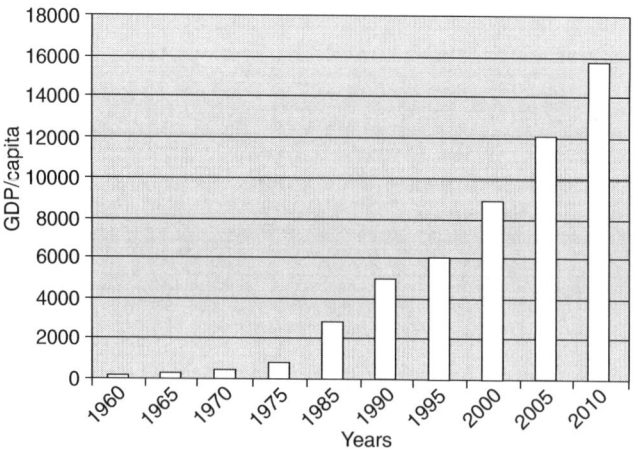

Figure 5. GDP growth, 1960–2010.
Source: World Bank (2010).

remarkable in the context of sub-Saharan Africa (World Bank 2010). With a Gini index of about 0.573, the government has aimed to ensure an equitable distribution of incomes through various programmes, regardless of where they live or what sector they work in, and with particular attention to economically empowering women and youth.

Successive surveys of income poverty show that the percentage of households and individuals with incomes below the poverty line has continued to fall (NDP10, 40). In addition, access to public services, notably education, health services and clean water, rose (NDP10, 40).

The government is implementing a long-term Development Plan known as 'Vision 2016' aimed at realisation of the United Nations Millennium Development Goals. Most of the funding is derived from the mining sector. It has maintained a variety of safety nets to cater for the vulnerable groups encompassing the elderly, youth, rural women and HIV/AIDS victims and orphans. These programmes include a universal non-contributory old age pension scheme, destitute allowances, the World War II Veterans' Programme, the Primary School Feeding Programme, the Vulnerable Group Feeding Programme and the National Gender Programme grant. There is good reason to believe that economic growth and government welfare programmes will generate further decreases in the number of people defined as living in poverty. The expected decline in diamond revenue, after the end of NDP 10, could however make it more difficult to maintain government expenditure on safety nets.

In mining settlements, mine employees are paid family wages rather than the single bachelor wages associated with the Southern African migrant labour system of the apartheid era. Estimates are that the 2600 workers at Orapa–Letlhakane mines directly sustain 10,000 dependants and the 5000 employees at Selebi Phikwe help to sustain township inhabitants numbering 44,600 while 2320 employees at Jwaneng mine support 14,000 dependants (Government of Botswana 2001). Because of urban dwellers' strong links with their kin, urban and mining wage benefits have filtered to the rural areas for poverty alleviation. In general, migrants continue to maintain their links with their home areas through remittances, investments and periodic participation in the cultural affairs of their home areas. Upon retirement some return to their areas of origin.

Because of the volatility of the mining sector, the government has embarked on economic diversification efforts that involve manufacturing. Most of these have failed because of weak political commitment, especially with regard to supporting innovation, research and development. The government has had a slow rate of response towards addressing factors that inhibit private sector growth and competitiveness, including utility costs, cost of capital, business registration and licensing, delays in processing work and residence permits and access to land. Furthermore, there has been a proliferation of fragmented, uncoordinated policies and strategies, which have seriously undermined implementation and monitoring and accountability in the case of institutions (Sekwati 2010). Unless the latest economic diversification drive (EDD) overcomes these hurdles, it is unlikely to succeed where other attempts failed (Government of Botswana 2012).

The downside: large-scale mining's employment impasse

From 1972 to 2008, total employment increased from below 40,000 to approximately 316,000. In 1972, mining employed 9% of the total formal employees. Interestingly, despite the diamond boom that followed, the percentage has not changed. Other sectors, particularly wholesale and retailing, experienced some increase due to the rise in disposable incomes and consumerism. The increase in manufacturing, finance and banking reflects government attempts to diversify the economy from a resource-based one.

Mining has generated significant opportunity costs for the rural sectors of the economy. Selective migration of the economically active 20–34-year-old males draws human capital from their hinterland with the very young and older men and women primarily left in their villages of origin (Ntsebane 2010). The ratio of females to males is normally low for mining towns and this encourages commercial sex, a risky occupation associated with the transmission of AIDS. Meanwhile, in open access mining towns like Selebi Phikwe, uncontrolled rapid rural–urban migration has been responsible for the growth of squatter settlements.

In essence, labour supply is exceeding demand despite mine sector expansion over the past 30 years and numerous employment-generating policies implemented by the state. Botswana's large-scale mining has simply been unable to keep pace with the numbers of Batswana leaving the countryside in search of non-agricultural income-earning opportunities.

Conclusion

Botswana represents an all-too-rare example of a sub-Saharan African country that has used its mineral revenues to successfully foster development. This reflects the role of a developmental state that has evolved the institutional capacity to promote and coordinate economic development through a suite of industrial, planning and welfare policies (Weiss 2000). In particular, the state has consciously assumed a central role in shaping mining policies to support urban infrastructural and service development and poverty reduction. Their underlying aim has been to ensure that mineral development infrastructure confers benefits to the national population as a whole, avoiding the opening up of regional or rural–urban imbalances.

Economic indicators suggest that the country has managed to mostly steer clear of the dilemmas of the natural resource curse that plagues several mineral-producing countries worldwide in terms of poor governance, lack of democracy and gross welfare inequalities (Bebbington et al. 2008). In fact, the World Economic Forum ranked Botswana number one for good governance while the World Economic Forum Global Competitiveness Report 2010–2011 notes that among the country's strengths are its reliable and legitimate institutions (32nd), the efficiency of government spending (15th), public trust of politicians (21st) and judicial independence (30th). Transparency International has ranked Botswana as the least corrupt among African countries for the last several years (Transparency International 2007). However, opposition parties within Botswana feel that there is increasing state authoritarianism. These sentiments have recently been expressed in claims that the government has become bureaucratically centralised and less transparent. Furthermore, the country still faces problems of socio-economic inequality (Jefferis and Kelly 1999; Taylor 2003). Some of the contradictions of Botswana's development, including poverty amidst plenty, are rooted in the capitalist system that the country has followed (Tsie 1996, 15).

Nonetheless, prudent economic management, a buoyant economy, honest bureaucracy, democracy and political stability have attracted substantial foreign investment to the country. The development plans of Botswana have continually been based upon four principles: democracy, development, self-reliance and unity. Economic development has been accompanied by expanded employment opportunities for locals and expatriates in the public and private sectors. As a result, Botswana's poverty alleviation campaign has been highly successful. Since national independence, the Botswana population has transformed from primarily poor pastoralists to urbanites living in a medium-income country enjoying an average annual per capita GDP of almost US$16,000.

It is readily acknowledged that diamond mining has been central to this achievement and that diamonds in Botswana are not forever. There is a finite supply that is well on its way to depletion. The mining sector currently accounts for over 70% of national export earnings and nearly 50% of government revenue but this cannot continue. Thus, the government has not only planned for the equitable geographical spread of welfare derived from the country's mineral wealth. Through urban planning, diversification of economic activity into services and manufacturing has been heavily subsidised, preparing for the day when the financial symbiosis between mining and urbanisation ceases.

Notes

1. The government, however, believes small-scale mining in industrial minerals could enhance import substitution for construction materials and would include small-scale mining of limestone, clay, sand and gravel.
2. For economic and political reasons, South Africa has been trying to stabilise its mining workforce by offering better working conditions to nationals in terms of wages, housing, career advancement and longer contracts. Consequently, the number of seasoned mine labour recruits from Botswana dropped from 40,390 in 1976 to 19,648 in 1986 (Taylor 1990).

Note on contributor

Thando Gwebu is a Professor in the Department of Environmental Science at the University of Botswana. He teaches courses in population and the environment, human settlement and sustainable development. His recent research publications include: Public participation and urban rehabilitation in sub-Saharan Africa: A case in Botswana. *Ekistics*, in press (with K. Jain); Towards a theoretical exploration of the differential urbanization model: The Botswana case. *Journal of Social and Economic Geography* 97 (2006): 418–33; Development of Jwaneng Mine: Economic impacts and environmental implications. *Botswana Journal of Southern African Studies* 24 (2010): 159–79 (with I.P. Ntsabane, M.B.K. Darkoh, T.D. Gwebu and O. Otlogetswe Totolo). He can be contacted at: gwebutd@mopipi.ub.bw

References

Asare, B.K. 1999. Perceptions of socio-economic and environmental impacts of mining in Botswana: A case study of the Selebi-Phikwe copper-nickel mine. MSc thesis, Department of Environmental Science, University of Botswana.

Asare, B.K., and M.B.K. Darkoh. 2001. Socio-economic and environmental impacts of mining in Botswana: A case study of the Selebi-Phikwe copper-nickel mine. *East Africa Social Science Research Review* XVII: 1–41.

Basdevant, O. 2008. Are diamonds forever? Using the permanent income hypothesis to analyze Botswana's reliance on diamond revenue. *IMF Working Paper* WP/08/80.

Bebbington, A., L. Hinojosa, D.H. Bebbington, M.L. Burneo, and X. Warnaars. 2008. Contention and ambiguity: Mining and the possibilities of development. *Development and Change* 39: 887–914.

Colclough, C., and S. McCarthy. 1980. *The political economy of Botswana*. Oxford: Oxford University Press.

Gaolathe, B. 1997. Development of Botswana's mineral sector. In *Aspects of the Botswana economy*, ed. J.S. Salkin, D. Mpabanga, D. Cowan, J. Selwe, and M. Wright, 401–31. Oxford: James Currey.

Government of Botswana. 2001. *Botswana atlas*. Gaborone, Botswana: Department of Surveys and Mapping.

Government of Botswana. 2012. *Economic diversification drive*. Gaborone: Ministry of Trade and Industry.

Gwebu, T.D. 1987. Spatial organisation of urban centers in Botswana: The evolution of development with equity. *African Urban Quarterly* 2: 149–60.

Gwebu, T.D. 2006. Towards a theoretical exploration of the differential urbanisation model: The Botswana case. *Journal of Social and Economic Geography* 97: 418–33.

Harvey, C., and S.R. Lewis. 1990. *Policy choice and development performance in Botswana*. London: Macmillan Press.

Jefferis, K., and T.F. Kelly. 1999. Botswana: Poverty amid plenty. *Oxford Development Studies* 27, no. 2: 211–31.

Kontuly, T., and H. Geyer. 2003. Testing the differential urbanisation model in developed and less developed countries. *Tijdschrift voor Economische en Sociale Geografie* 94: 3–10.

Matshediso, I.B. 2005. A review of mineral development and investment policies of Botswana. *Resources Policy* 30: 203–7.

Ntsabane, I.P., M.B.K. Darkoh, T.D. Gwebu, and O. Totolo. 2010. Development of Jwaneng Mine: Economic impacts and environmental implications. *Botswana Journal of Southern African Studies* 24: 159–79.

Perroux, F. 1970. More on the concept of growth poles. In *Regional economics*, ed. D. McKee, R. Dean, and W. Leahy, 93–104. New York: The Free Press.

Republic of Botswana. 2001. *Census Population of and Housing*. Gaborone, Botswana: Government Printer.

Republic of Botswana. 2003. *National Development Plan 9*. Gaborone, Botswana: Ministry of Finance and Development Planning.

Republic of Botswana. 2009. *National Development Plan 10*. Gaborone, Botswana: Ministry of Finance and Development Planning.

Sekwati, L. 2010. Botswana: A note on Economic Diversification. *Botswana Journal of Economics* 5: 78–85.

Silitshena, R.M.K., and G. McLeod. 1998. *Botswana: A physical, social and economic geography.* Gaborone: Longman Botswana.

Taylor, J. 1990. The migration element in the 1981 Botswana census. *Botswana Notes and Records* 17: 89–91.

Taylor, I. 2003. Ditiro tsa ditlhabololo: Botswana as a developmental state. *Pula Botswana Journal of African Studies* 17: 37–50.

Taylor, I., and G. Mokhawa. 2003. Not for ever: Botswana, conflict diamonds and the Bushmen. *African Affairs* 102: 261–83.

Transparency International. 2007. Corruption perceptions index. http://www.democracyweb. org/accountability/botswana.

Tsie, B. 1996. The political context of Botswana's development performance. *Journal of Southern African Studies* 22: 599–616.

Weiss, L. 2000. Developmental states in transition: Adapting, dismantling, innovating, not 'normalising'. *The Pacific Review* 13: 21–55.

World Bank. 2010. *World development indicators.* http://data.worldbank.org/data-catalog/ world-development-indicators.

Zelinsky, W. 1971. The hypothesis of the mobility transition. *Geographical Review* 61: 219–49.

Unearthing treasure and trouble: mining as an impetus to urbanisation in Tanzania

Deborah Fahy Bryceson, Jesper Bosse Jønsson, Crispin Kinabo and
Mike Shand

Despite an abundance of mineral wealth and an ancient history of gold trading, Tanzania is a relative latecomer to the experience of being a mineral-dominated national economy. Both the British colonial state and Nyerere's post-colonial state avoided encouraging, and only reluctantly provided support to, large- and small-scale mining. Farming constituted the livelihood for the vast majority of the population and peasant agricultural exports provided the main source of foreign exchange for the country. Now, however, Tanzania has become one of Africa's main gold producers and the number one destination for non-oil foreign direct investment after South Africa. This article traces the development of gold mining and urban growth in Tanzania with the aim of identifying if, when and where these two processes interact with one another. It explores the triggers, mechanisms and durability of their fusion and synergies over time.

The search for a better life has propelled millions of Tanzanians to migrate from rural homelands to urban areas since national independence in 1961. With the steady erosion of peasant farmers' terms of trade over the last 30 years, members of farming households have been re-evaluating their links to the land (Bryceson 2009). Increasingly many have come to see the countryside as a poverty trap from which they seek escape. Formerly, most were attracted to the capital city, or nearby secondary cities. Now many migrate to mining areas, usually located in remote rural areas. Over time, however, the pressure of population amassing at the mineral-bearing locality is likely to generate urban growth. Mineral depletion can set in train a converse process of depopulation and impoverishment. The first to be affected are the small-scale miners[1] and their dependants, with reverberating impacts on service providers.

In this contribution, we trace mining and urbanisation's impact on the Tanzanian population and economy with respect to urban population densification, economic diversification and social stratification. Our aim is to identify where and when mineral production and trade in Tanzania catalysed interactive processes and synergies between mining and urbanisation, paying particular attention to the intensity of the synergy affected by the high monetary value of the extracted minerals destined for global markets relative to the low level of economic welfare of the majority of the national population. Conversely, mining and urbanisation synergies are anticipated to weaken as the mineral supply dissipates and/or the market value of the minerals declines.

By deploying the term synergy we are not inferring that the interactive processes are simple, formulaic straight-line paths to urbanisation. Far from it, the dynamics are complex and liable to instability and reversal due to internal imbalances or negative external stimuli as inferred in the juxtaposing of 'treasure' and 'trouble' in the title of this contribution. The welfare of urban residents is most at risk. Migrants are attracted to mining settlements with expectations of material gain, but the reality is that people necessarily lead highly competitive, risky lives. Access to livelihoods, services and basic social needs within the urban community are uncertain.

Large-scale mining corporations' investments are likely to catalyse urban growth in developing country contexts often preceded by activities pioneered by small-scale miners and other local agents. The latter's migration to mineral-rich localities is energised by the implicit hope of narrowing the gap between perceived local and global economic differences. In the case of gold such aspirations are not unreasonable. The world price of gold has climbed to an unprecedented level, at the same time as small-scale miners have been able to keep themselves informed of world prices and use that information to bargain for better gold prices.

Our objective is to probe urbanisation in relation to small- and large-scale mining in Tanzania. The first section focuses on mining history, notably the peaks of urban growth and mining activities of the thirteenth–fourteenth centuries when Kilwa Kisiwani, an island off the Tanzanian coast, was famed for its urban opulence and splendour as the gold entrepôt of East Africa. Thereafter our historical review moves forward in time to Tanzanian mining developments in the nineteenth and twentieth centuries, specifically at two gold-producing settlements, Lupa and Geita. Tracing Tanzania's entwined mine-led migration and urbanisation history, the country's urban growth pattern and the resultant urban hierarchy are discussed in light of Geyer and Kuntuly's (1993) concept of depolarisation and differential urbanisation. Turning to the significance of settlements with secondary connections to gold mining, we interrogate how service-provisioning activities form part of the geographical spread of mining's urbanising influences and provide possible exemplars for mining towns facing mineral depletion. Our conclusion raises questions regarding the sustainability of individual mine-led urban settlements and Tanzania's urban future more generally.

Kilwa: gold-rich global city of the thirteenth–fourteenth centuries

Kilwa's fame as the East African coast's main trading centre for the export of Zimbabwean gold is linked to the emergence of an extremely cosmopolitan form of urbanism in East Africa at the beginning of the twelfth century, peaking between 1250 and 1350, before ebbing away with dwindling gold production in Zimbabwe. None of the gold trade was known to be sourced from Tanzania. Rather it was mined and transported from Zimbabwe and Manicaland overland to Sofala in Mozambique, then shipped northwards in small boats up the Swahili coast to Kilwa, where it was loaded onto larger vessels destined for the Middle Eastern and Indian gold markets. Kilwa's trans-shipment harbour was strategically situated for larger sea vessels travelling to and from India which relied on seasonal monsoon winds. Further south along the Mozambican coastline, the winds were too weak for the large sea-going vessels (Pearson 1998).

In a period when European cities were just beginning to distinguish themselves from Europe's feudal countryside, Kilwa was in effect a global city of the medieval period. The Madhali dynasty built the domed extension of the Great Mosque and the enormous palace and extensive trade depot of Husuni Kubwa in the early fourteenth century, which is unsurpassed in coastal Swahili architecture, incorporating many exotic Middle Eastern design features found elsewhere in the Islamic world (Mitchell 2005). Its construction is estimated to have taken place between 1315 and 1330 when the world price of gold was peaking under the sultanate of Al-Hassan bin Sulaiman (Moon 2005). At the time, the explorer Ibn Battuta (1962) remarked that Kilwa ranked as one of the most well-constructed and visually pleasing towns in the world.

Nonetheless, at its peak of urban sophistication, Kilwa became subject to the fluctuation of gold prices. The Black Death caused a sudden, sharp fall in the world gold price in Europe and beyond in the 1340s. While demand waned, the supply of Zimbabwean gold was diminishing, undermining Kilwa's gold economy and prosperity, as suggested by the prolonged state of disrepair of the dome in the Great Mosque (Moon 2005). Many decades passed before the Great Mosque's dome was repaired and new building works were started in the early fifteenth century. Foreign visitors were again impressed in the sixteenth century (Pearson 1998), noting the extravagance of the jewellery and clothing of the stone town's elite population.

At the time of the arrival of the Portuguese in East Africa in 1498, Kilwa was estimated to have a population of 5000 to 10,000 people. Already there were outward signs of declining wealth and waning trading dominance.[2] The Portuguese seized Kilwa in 1505 and built a fort (*gereza*) but relinquished the island in 1512, favouring Mombasa as their East African naval base.[3]

Today, Kilwa has a very small village settlement mostly of fishermen. Its majestic ruins attract foreign tourists in small numbers but its connection to an international gold trade is largely forgotten. The significance of gold does not reappear in Tanzanian history for almost four centuries.

Mining in the nineteenth and twentieth centuries

Since the colonial intrusion in 1888 right up to the 1990s, Tanzania's economic potential was primarily perceived in terms of its surface land, soils, water resources, and wildlife. Agriculture was the economic foundation of the country, with tourism only gaining significance during the 1980s and 1990s as peasant agriculture declined. The country's mineral resources – gold, diamonds and a variety of precious stones and industrial minerals – were not taken into account or planned for, despite the country's economic poverty.

Nonetheless, a century before, German colonists' mineral prospecting had revealed mica, copper, iron, lead, iron and zinc at various locations in the territory. In 1898 the German company, Koncession für Edelmineralien, discovered gold at Saragura Hill in the north-western part of the country, 15 kilometres from where Geita town is situated today (Chachage 1995). Shallow-level production was initiated, but as gold prices were low and World War I loomed on the horizon, they did not proceed to deep-level mining (Lemelle 1986). Similarly, gold was discovered at Sekenke in 1906 and production began in 1909. The following year nearly a ton of gold was extracted from the mine, with the workforce expanding to 20 European and 700 locally recruited workers soon thereafter.

German East Africa became the site of intense armed conflict between German and British-led troops during World War I, putting a stop to all forms of mining. In the transition from German to British rule, there was a hiatus in titled property ownership. Holding a League of Nations mandate to administer the territory, the colonial state, which was legally obliged to give precedence to African interests over those of non-indigenous settlers, avoided encouraging or supporting European mining initiatives. Geo-politically, the tense inter-war period, in which German imperial interests re-emerged under Hitler's leadership, left potential venture capitalists doubtful about the security of their investments in Tanganyika (Lemelle 1986). Thus, a significant amount of the mining activity that took place during the 1920s and 1930s under British rule arose in the context of declining agricultural prices and soaring gold prices during a world depression.

Lupa, Geita and beyond: mine rush towns

The discovery of gold at Lupa in 1922 in a remote part of south-western Tanganyika triggered the territory's first gold rush. Lupa was an arid, uninhabited and inaccessible plain in the mid-1920s, when the gold rush began. From a mere trickle of non-native diggers and their African labour, the Lupa gold field grew to a population of more than 20,000 by the mid-1930s. The number of Europeans expanded from 150 in 1922 to over 1000 in 1936, with the African diggers, many of whom came from the neighbouring countries of Malawi, Kenya and Zambia, and some as far away as Zimbabwe, being in the majority (Dickson 1925; Lemelle 1986; Bryceson 1990). By the early 1930s, Lupa had gained the reputation of being Tanganyika's urban slum. With diggers and labourers continuing to pour in, sanitation and food supplies were stretched far beyond the limits of public health. Scurvy was rife and the prices of basic foodstuffs were exorbitant. Despite deplorable conditions at Lupa, the government was slow to take action amidst the Depression. In 1935, a road was built to facilitate the transport of food. The Lupa gold rush had a magnetic force on the settlement pattern of Mbeya region. Men attracted to the gold fields for mining were followed by others eager to provide tertiary services and foodstuffs (Roberts 1986). Official gold production in Tanzania went up from 467 kilos in 1928 to 1711 kilos in 1934 (Lemelle 1986, 125).

Mining in the Lupa gold fields was dominated by alluvial mining until 1939 when reef mining started exceeding alluvial mining output (Roberts 1986). At the outset of the rush, gangs of young Africans worked for their white bosses. However, the 1929 Mining Ordinance gave all those who could afford 10 shillings and could fill out a form the possibility to apply for a prospecting licence. Thus, from the 1930s African miners started to access prospecting rights themselves (Lemelle 1986).

Large-scale production from the reef mines around Lake Victoria, predominantly at Geita, Musoma and Sekenke, gained pace from around 1930. Geita Mine became the largest gold producer in the region in 1936. Settlement conditions for the African miners were better there. Geita Mine, in particular, was officially viewed as a model workplace with respect to its provision of a balanced daily diet of cooked food for the workers and living conditions (Tanganyika Territory 1938; Thompson 1936).

The late 1930s were golden years for the Tanganyikan colonial mining sector. Gold production peaked at four tonnes in 1939, largely due to high world market prices that prevailed during the global recession. Gold exports became a significant contributor to the territory's economy, constituting almost 20% of export earnings in the late 1930s,[4] creating employment for and tax income from miners. In 1938, a committee under the colonial administration estimated that there were 32,000 Africans involved with mining in Tanganyika, 85% in gold, with the majority (73%) working as small-scale miners in Lupa (Lemelle 1986).[5] Already in those early years it was clear that small-scale mining was more labour absorbing than large-scale mining.

During World War II, prospecting was banned (from 1941) and many mines first cut down on their production and subsequently closed down altogether due to labour shortages and difficulties in accessing supplies (Roberts 1986). When the gold mines in Tanganyika resumed production after the war it was at a lower scale, with annual production at two tonnes and mineral exports constituting only 5% of export earnings. In the late 1940s and into the 1950s many gold mines closed and the ones that continued to produce gold, especially Geita Mine, faced financial problems. Most other small gold mines, in Chunya, Mpanda, Buhemba, Kahama and Nzega districts, also struggled financially during the 1950s (Chachage 1995).

In 1961 Tanzania attained national independence. Despite gold, diamond and gemstone discoveries during the colonial period, President Nyerere, as father of the nation, was reluctant to pursue exploitation of the country's mineral wealth or see minerals as foundational to national economic growth. During the first years after independence, various mines continued production. At the same time new discoveries of gemstones (e.g. tanzanite, ruby, sapphire, alexandrite and emerald) and gold were made at several locations. However, with the closure of Geita Mine in 1966, large-scale gold mining ended. From 1967, Tanzania embarked on a path of central planning and nationalisation of most of the country's mines to form a state-controlled mining sector run by the state-owned mining corporation STAMICO. Official gold production declined from three and a half tonnes in the mid-1960s to 10 kilos in the early 1970s. This, however, masks the fact that many of the Tanzanians who had been employed in gold and gemstone mines continued to mine with rudimentary tools, and smuggled their produce out of the country, mainly to Nairobi in Kenya. Small-scale miners' activities during the 1960s and 1970s were clandestine, taking place beyond the control of STAMICO in remote parts of the country (Drechsler 2001). At the end of the 1970s, Tanzania's mining sector consisted of a few dysfunctional industrial mines and scattered small-scale miners numbering probably somewhere between 10,000 and 50,000.

National gold rush momentum

With the promulgation of Tanzania's first post-independence mining legislation, the Mining Act of 1979, the Government of Tanzania abandoned its insistence on mining being controlled exclusively by the state (Kulindwa et al. 2003). The Act gave citizens the opportunity to post mining claims in areas designated for prospecting and mining for minerals and engage in mining activities that did not require large expenditures and specialised equipment. Large-scale mining operations, however, were still only accepted if carried out in partnership with state enterprises (Jønsson and Fold 2009). The Act was followed by the Small Scale Mining Policy Paper of 1983, which encouraged Tanzanians to supplement their incomes by participating in mining activities (Chachage 1995).

In the 1980s, IMF-imposed structural adjustment coupled with prevailing low agricultural commodity prices increasingly undermined the viability of the peasant agricultural sector (Bryceson 1999). Rural households coped with adversities by looking for alternative non-farm income sources. Small-scale mining provided an alternative to agriculture in mineral-rich parts of the country (Chachage 1995; Mwaipopo et al. 2004; Bryceson and Jønsson 2010). The 1980s saw a rapid succession of mineral discoveries of gold and gemstones. The fact that mineral smuggling was rampant is indicated by the meteoric rise in official gold exports after the breakup of the STAMICO monopoly at the end of the 1980s. Thereafter the government commissioned two national banks to buy gold from anyone at close to world market prices. The international gold price was on an upward trajectory. From virtually zero, Tanzania's official gold exports increased to four and a half tonnes in 1992 (Kulindwa et al. 2003; Phillips et al. 2001).

The 1990s and 2000s saw numerous gold and gemstone rushes and small-scale mining sites emerged throughout Tanzania (Figure 1). Geita district, in particular, experienced extremely rapid population expansion, doubling between the censuses of 1988 and 2002 with in-migration of people attracted by the opportunities of small and large-scale mining, especially Tanzania's biggest gold mine Geita Gold Mine owned by AngloGold Ashanti, to become the most populous district outside of Dar es Salaam (Lange 2006). Recently it has received regional status in recognition of the density of its population and its contribution to the economy.

Throughout the 1990s, hundreds of thousands of small-scale miners increasingly existed alongside western commercial exploration companies. The presence of the latter was triggered by a set of World Bank-instigated investment and mining law reforms endorsed in the late 1990s containing the key elements required to attract foreign direct investment (FDI), i.e. the 1997 Mineral Policy, the 1997 Investment Act, the 1998 Mining Act and the 1999 Mining Regulations. As intended, the new legislative framework was exceptionally effective in pulling in foreign direct investment, which since 1998 has amounted to more than US$2.5 billion in the Tanzanian mining sector, especially in gold exploration and mining.

Besides a modern gemstone mine in the north-eastern part of the country, all large-scale mines (six gold mines and one diamond mine) are located in the north-west around Lake Victoria in the regions of Tabora, Shinyanga, Mwanza, Geita and Mara. Observing the number of gold sites ringed around the Tabora region in Figure 1, we will refer to these as Tanzania's 'ring of gold' regions. In 2009, gold exports alone earned close to US$1.4 billion. In only 10 years, Tanzania has become one of the main African gold producers with annual outputs of around 50 tonnes (Roe and Essex 2009; Emel and Huber 2008).

News of a gold strike travels quickly with the increasingly ubiquitous use of mobile phones, generating a flash flood phenomenon as thousands of miners rush to the strike site to get a chance to work claims with the richest seams of ore (Jønsson and Bryceson 2009). Over the last two decades, miners have had many mineral sites to choose between and the numbers of small-scale miners have proliferated. In the mid-1990s, the number was estimated to be 550,000 (Tan Discovery 1996) and a recent World Bank-funded study conducted in 2011 estimates the number to be

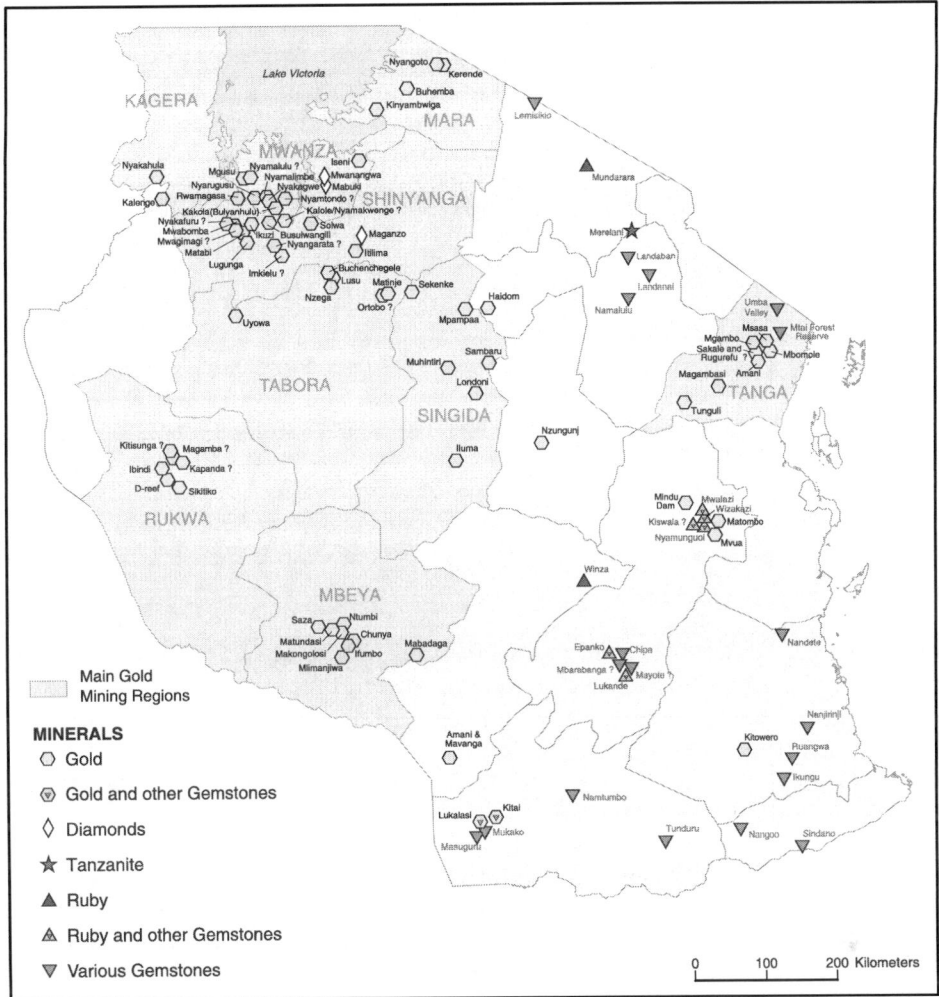

Figure 1. Tanzania small-scale mining sites.
Source: Shand and Jønsson (2011) UPIMA project compiled from Jønsson's Tanzanian small-scale mine list.

685,000.[6] As a result Tanzania's government recognises that the small-scale mining sector is a major provider of economic opportunities to rural communities (Tanzania 2005). In comparison, between 2003 and 2009, the number of employees within the large-scale mining sector varied between 6000 and 12,000 (Roe and Essex 2009). These employees, however, earned significantly more than the average small-scale miner.

Small- versus large-scale mining

Following the endorsement of the 1998 Mining Act, an unprecedented debate on the mining sector in Tanzania began. Critics emphasised the limited revenues that the Tanzanian state received from minerals, especially gold, exported by large-scale mining companies, the often insensitive evictions (or resettlement schemes depending on the observers) of small-scale miners and farmers brought about by the same companies, and the fact that the legislation benefited large-scale mining at the expense of small-scale miners. With the 1998 Mining Act small-scale miners had the same opportunity to secure prospecting and mining licences as medium- and large-scale companies. However, the restricted capacity of mining authorities to disseminate information on claim availability to small-scale miners as well as how and when to acquire licences favoured professionalised medium- and large-scale mining companies which could secure licences in mineral-rich areas before the majority of small-scale miners knew of the opportunity (Jønsson and Fold 2009; Carstens and Hilson 2009).

To address the criticism, President Kikwete set up a Presidential Mining Review Committee in 2007, which emphasised the need for an enhancement of the state's share of production from large-scale mining. This was to be done through increased taxation and royalties and equity share ownership and also through improved working conditions for small-scale miners (Tanzania 2008). An immediate consequence of the report was a planned revision and reform of the mining legislation.

In July 2009, Tanzania launched a new Mineral Policy aimed at promoting increased integration between the mining sector and other sectors of the economy in order to improve mining's contribution to the national economy.[7] The policy also prescribed the development and increased formalisation of the small-scale mining sector in order to facilitate sustainable development (Tanzania 2009).

In April 2010, parliament passed the 2010 Mining Act (Tanzania 2010) with the aim of achieving more equitable production and distribution arrangements for the financial proceeds of mining to alleviate voter dissatisfaction with the mining industry and to secure a continuous in-flow of FDI into the mining sector. The legislation designated more areas exclusively for small-scale mining purposes and imposed royalties on large-scale metal production up from 3% to 4% Large-scale mining companies have to be listed on the Dar es Salaam Stock Exchange, and the government has the possibility to acquire stakes in all future mining projects. Despite the new provisions towards small-scale mining, however, essentially the reformed legislation is still mainly targeting large-scale mining companies through an investor-friendly environment (Jønsson and Fold 2011). Thus, in Tanzania today, small-scale miners and large-scale mining companies exist side-by-side, though not hand-in-hand.

Migration to mining settlements

While many who gravitate to small-scale mining settlements go bankrupt or lose their motivation given the physical danger associated with mining and the financial insecurity of not knowing if and when minerals will be found, there are nonetheless enough miners who do succeed to continually fuel people's expectations of lucrative earnings and replenish the migration flow (Jønsson and Bryceson 2009). Unlike the well-documented process of migration to towns for formal jobs or more latterly informal sector work, which transports individuals into urban lifestyles (Todaro 1969, 1997; Beall et al. 1999; Potts 2010), the rapid urbanisation of mining settlements affords individuals the experience of being catalysing agents for *in situ* urban birth and transformation. The first and subsequent migrants arrive in a rural setting that is taking on urban features as population concentration in the settlement outgrows reliance on local natural resources like water and firewood and starts contending with decreasing availability of land for food production (Bryceson and Mwaipopo 2010; Bryceson and Jønsson 2010).

Despite their rough and ready appearance, mining sites are usually quite cosmopolitan in character, attracting urban- as well as rural-born migrants from close as well as very distant places.[8] Muzzini and Lindeboom's (2008, 52)[9] analysis of 2001–2002 data reveals that migration activity was most pronounced in the central mining core regions, Tanzania's 'ring of gold', where in-migration to regional towns generally superseded migration to the regional capitals.[10] Interestingly, Shinyanga's regional capital in fact recorded net out-migration, as did Singida's capital to a lesser extent. Both Singida and Tabora distinguished themselves as the only western regions that experienced net urban out-migration, which may reflect the lower quantity and quality of their mineral strike sites relative to the other mining regions (Muzzini and Lindeboom 2008).

What makes Tanzania's mine-led migration profoundly urban rather than rural in character is that the migration stream is not limited to mine labourers. It is highly diverse in skills and geographical origin with the potential to create a multi-occupational urban population from the outset. In the first instance, it is predominately male-biased as men rush to the mineral strike site, but within weeks women as well as many men are following with the intention of working in trade and service sector activities. Subsequent waves of non-mining migrants are responding to the same stimulus as miners. They anticipate higher earnings due to a sharp rise in demand for service activities on the part of miners with purchasing power far in excess of farmers in the surrounding rural countryside. Some women may be following husbands, but very large numbers will be part of chain migration through friendship networks, rather than family ties, seeking an independent income (Bryceson, Jønsson and Verbrugge 2013 forthcoming). Thus, the settlement grows with women's active involvement in tertiary services as well as mineral processing work.

Having discussed the growth of individual mining camps from a rural to urban scale of settlement, we will now consider Tanzania's unfolding pattern of urbanisation in order to identify when and where the impact of mining surfaces at a national level.

Post-independence urbanisation trajectory

At national independence in 1961 Tanzania was an overwhelmingly rural country (Bryceson 1990). Only 7% of the population lived in urban areas and Dar es Salaam dominated the urban hierarchy. As the de facto colour bar was lifted and white- and blue-collar urban jobs became available to Africans, young men as well as women gravitated to the city. Dar es Salaam, the nation's capital, was their favoured destination. Hirst (1972) provided a perceptive analysis of Tanzania's urban hierarchy in 1969 when Dar es Salaam was at its zenith as the country's primate city in the first decade after national independence. The city's dominance in the nascent urban hierarchy became a government concern. By the 1970s, policymakers held the view that the capital's employment opportunities should not be allowed to expand out of proportion to the rest of the country. The implementation of a decentralisation policy which involved posting civil servants and party officials to the capitals of mainland Tanzania's 20 regions, as well as a growth pole strategy, aimed at spatially dispersing Tanzania's embryonic industries to seven cities (Nyerere 1972; Bryceson 1984). The latter did not achieve much success because, unlike the port city Dar es Salaam, Tanzania's other towns were constrained by costs of long-distance goods transport to and from the coast and did not have the critical economics of scale and infrastructure that Dar es Salaam offered. However, government decentralisation of bureaucratic staff created nuclei of demand for urban goods and services that started to spur the growth of the country's 20 regional capitals, thereby reducing the urban primacy of Dar es Salaam (Figure 2).

A differentiated picture of urban growth emerged in the 1980s. On the whole, population growth slowed down in Dar es Salaam, the national capital, and accelerated in the regional urban capitals, as indicated by the steeper incline of the regional capitals' population growth between the 1967 and 1988 censuses (Figure 3).

However, according to Tanzanian census data, there were distinct growth rates between capitals of mining regions and non-mining regions. In the inter-censal period between 1978 and 2002, mining regional capitals grew at 3.5% per annum relative to non-mining capitals at 4.6% (Table 1). As small-scale mining spread, the growth of small towns in mining districts came to the fore. When one compares the growth rates of small towns of 10,000 or over with Dar es Salaam during the 1980s and 1990s, the small towns of mining regions are notable for their exceptionally high annual growth relative to all other settlement types. Dar es Salaam, non-mining regional capitals and small towns in non-mining regions were all growing at roughly the same pace (4.7%) while small towns in mining regions accelerated to 7.4%. There was a tendency for traders and service sector provisioners to readily move from their residences in the regional towns to nearby mineral strike sites in anticipation of higher income, giving rise to the rapid growth of small town mining sites at 7.4% per annum.

The urban literature to date has not focused on the changing pattern of urbanisation triggered by Tanzania's mining surge. This is understandable given the sudden and seemingly ephemeral nature of small-scale mining settlements, in

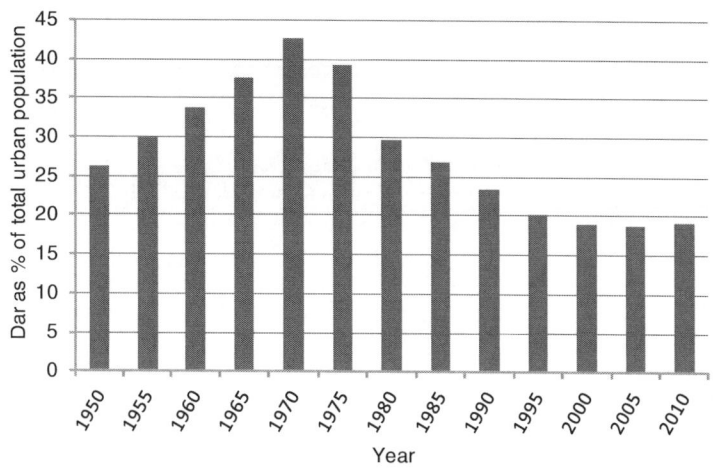

Figure 2. Dar es Salaam urban primacy, 1950–2010.
Source: United Nations, Population Division (2004).

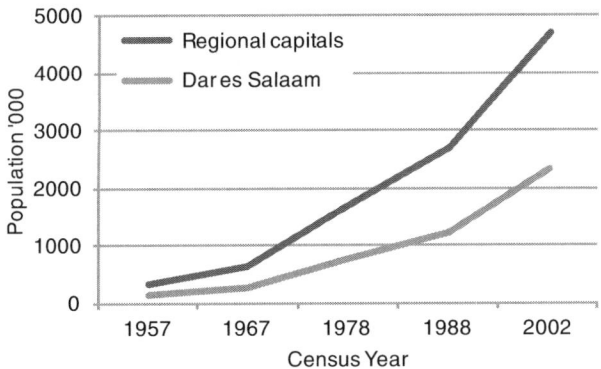

Figure 3. Urban population growth, 1957–2002.
Sources: Tanzania censuses 1957, 1967, 1978, 1988, 2002. (Tanganyika Territory 1963; Tanzania 1970, 1982, 1991, 2012).

addition to the fact that mine-led migration taking place between censuses goes largely unrecorded. To address this lacuna, we have combined the e-Geopolis data set[11] with an up-to-date list of small-scale mining sites to categorise urban settlements into 'mining' and 'non-mining' regions and districts (Shand and Jønsson 2011). In this way, it has been possible to generate maps showing Tanzanian urban settlement growth change in relation to the location of small-scale mining sites.

The results of the last two censuses hint that Tanzania may be evidencing what Geyer (1996) refers to as 'depolarisation' whereby core areas of urban growth relinquish their growth rate lead in favour of less populated locations. This involves 'differential urbanisation' in which fast growth followed by slower growth is experienced in successive waves that engulf groups of large, then intermediate-sized, and finally small cities. Geyer and Kontuly (1993) argue that this continuum characterises the evolution of urban systems in both developed and developing countries. By way of illustrating the process of depolarisation embedded in differential urbanisation, Dar es Salaam's urban primacy peaked in the early

Table 1. Annual urban growth rates by non-mining and mining regions and urban categories, 1978–2002.

Urban category	1978	2002	Annual growth rate (%)
Dar es Salaam	769,445	2,339,910	4.7
Non-mining regional capitals	424,614	1,246,059	4.6
Mining regional capitals	488,822	1,121,239	3.5
Towns in non-mining regions	252,634	784,840	4.8
Towns in mining regions	163,841	911,428	7.4

Notes: 'Mining region' in this categorisation refers to Tanzania's main gold and diamond mining regions (Mara, Mwanza, Shinyanga, Kagera, Singida, Tabora, Rukwa, Mbeya, Tanga), identified on the basis of being regions where 25% or more of the rural districts have known small-scale mining sites of gold or diamonds. Note Morogoro, Dodoma, Arusha, Manyara, Mtwara, Lindi and Ruvuma regions with mining sites of precious stones (emeralds, rubies, sapphires, etc.) achieve this level of prevalence in production of precious stones but not diamonds or gold and are therefore excluded from our mining region category. Source: Tanzania censuses 1978 and 2002. A breakdown of urban populations by region was not available for the 1988 census.

1970s, superseded by the rapid growth rates of Tanzania's regional capitals in the decade thereafter, followed by the emergence of rapid small town growth particularly in mining regions (Bryceson 2006). This expanding depolarisation process is sometimes confusingly referred to as 'counterurbanisation' when it refers to small town growth at the expense of primate city growth (Geyer 1996). We prefer to avoid using 'counterurbanisation' in this context, reserving it instead for instances of absolute urban depopulation.

Figure 4 juxtaposes urban settlement growth over 20-year intervals, revealing how urban population growth has been heavily concentrated in the mining regions, especially in the last 20 years, with an increasingly obvious 'ring of gold' pattern emerging in the western part of the country, reflecting the location of mineral sites.

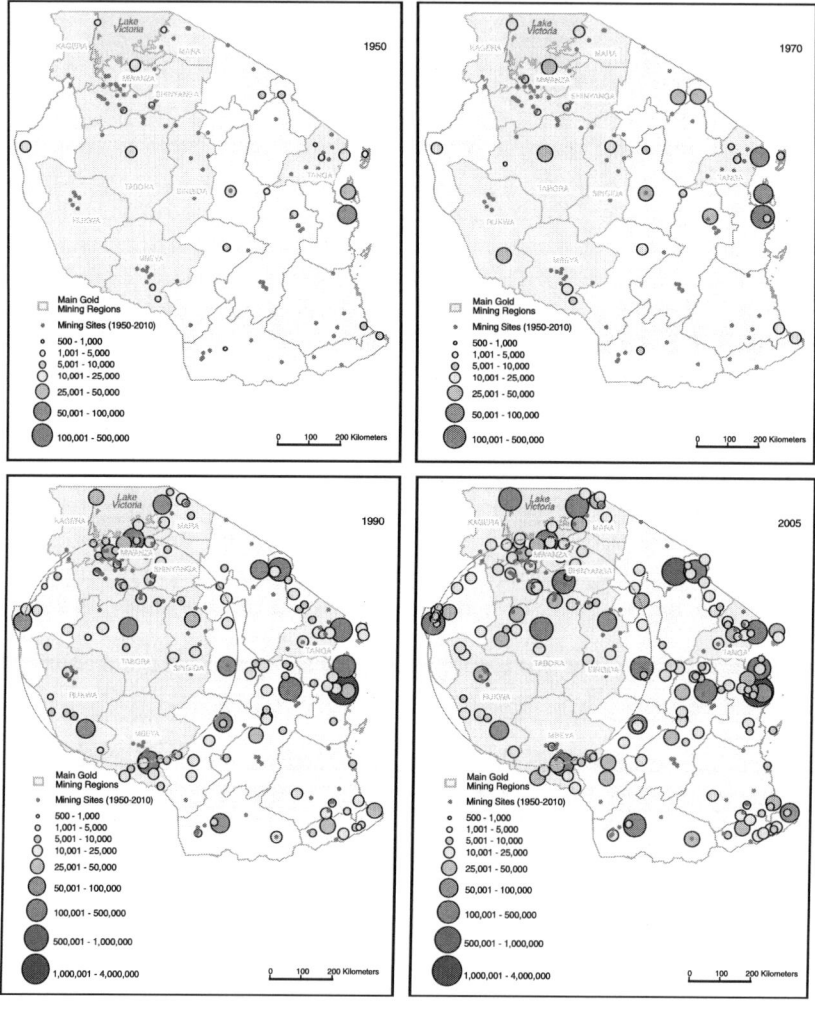

Figure 4. Tanzanian small-scale mine sites relative to urban growth, 1950–2005.
Source: Shand (2011) UPIMA project, compiled from Jønsson's Tanzanian small-scale mine list, e-Geopolis urban population data and the Second Administrative Level Boundaries data set project (SALB).

Comparing 1970 and 2005 in Figure 4 provides visual evidence of the clustering of urban settlements in gold-rich areas in Tanzania's north-western mining frontier, namely Mara, Mwanza and Shinyanga regions[12] (Bryceson and Mwaipopo 2010). Many of these areas are inhospitable, semi-arid areas prone to water shortages. Nonetheless, the lure of mineral strikes funnels the migration of often extremely large numbers to remote sites regardless of water availability and agricultural potential. While a comparison of the mineral site locations of Figure 1 and the urban settlement pattern of Figure 4 hint at a strong association between mining and population growth, further survey and census data would be required to pinpoint the strength of the association.[13]

Over time, in-migration is likely to taper and begin to be replaced by out-migration, particularly when small-scale miners reach a point of technological incapacity preventing them from digging deeper mining shafts or are evicted by the encroachment of large-scale mining. As male miners migrate elsewhere, they may leave behind a larger resident population than existed at the outset of the mining boom, including their wives, girlfriends and children. In the process, the gender ratio of the settlement is likely to tip from a male to a female bias. Beyond the changing gender ratio, counter-urbanisation in the form of absolute depopulation may take place in some settlements, with some eventually becoming 'ghost towns', abandoned for lack of viable livelihoods. In locations with good water supplies, fertile land and proper road connections, urban growth is more likely to be sustained, bolstered by government, civil society and private sector infrastructural investment in roads, schools, health clinics and religious centres.

In these cases, the settlement is in a position to draw additional migrants from rural areas, which are attracted to the superior amenities. This, however, depends on some degree of alternative employment prospects and/or the possibility of an agrarian fall-back. This outcome points to another pathway through which mineral wealth promotes urban growth, as illustrated by Katoro-Buseresere, a settlement of over 30,000 people (2002 census) located in the north-west section of Tanzania's ring of gold in Geita region, which has no direct mineral wealth, but instead is a thriving market town with an extensive service sector, that has attracted the custom of traders, transporters and small-scale miners (Bryceson 2011). Demographically, during the boom years of the 1990s its population grew very rapidly, biased towards female rather than male migrants. Overwhelmingly their motivation for migration was the search for trade and better opportunities rather than marriage or employment expectations. Its position on the main road to Uganda, straddling two regions, has given it additional commercial potential. A number of small-scale gold mining sites surrounding the town make it a favourite place for miners when they have cash to splash around. The livelihood patterns of the settlement range across trade and agriculture, giving it a far more stable foundation for growth compared with the fluctuating fortunes of pure mining settlements (Bryceson and Mwaipopo 2010).

Ring of gold promises and problems for Tanzania's urban future

Having traced Tanzania's mining history in tandem with its urban history, we have identified three mineral-related stimulants to urban growth: first, sites of regional or international trade in minerals, exemplified by Kilwa, a cosmopolitan city of the medieval past that thrived as an Indian Ocean gold trading centre. Second are locations of on-site mineral production as illustrated by the Lupa gold rush during the colonial period, and Geita, a large-scale corporate mining town of the present. Both feature strong population growth and economic diversification. Third are sites of secondary service development, notably Katoro-Buseresere, which rely on the increased purchasing power of regional mining wealth.

Tracing the processes of mineralisation and urbanisation, we have observed that colonial gold and diamond production in Tanganyika were not dominant forces in the territorial economy. However, during the world depression of the 1930s, as the price of gold rose, the country experienced a localized gold rush in Lupa that is incomparable with the magnitude of the country's present surge in gold production. A spiralling gold price amidst global economic uncertainty has spurred small-scale miners' discovery of a proliferation of gold sites, especially in Tanzania's ring of gold. Miners have rushed to these sites in the hopes of profiting. In the process, many remote rural locations have quite suddenly grown into sizeable small towns. Gold now constitutes more than 50% of the value of the nation's exports. Small-scale production of a clandestine nature has been giving way to more open small-scale mining that now competes with large-scale mining interests for mineral rights.

The analysis here has marshalled evidence to suggest that a synergy between mining and urban growth is currently underway, most heavily influenced, in the first instance, by small-scale miners' migration to mineral rush sites. The locational shift of an economically active population has catalysed rapid secondary city growth in the country's mineral-rich regions. Our analysis of Tanzanian census data indicates that urban growth has progressed from an early period of primate city dominance (up to the 1970s), through a period of rapid growth of intermediate secondary cities, notably the country's regional capitals during the 1980s, to a third stage in which the rapid growth rate and proliferation of small towns, specifically in mineral-rich areas, is now pronounced. The three-stage pattern of successive fast then slow waves of growth of the primate city, intermediate cities and finally small towns, identified by Geyer and Kontuly (1993), is reflected in the settlement growth pattern of Tanzania's urban hierarchy.

In addition to rapid population expansion, a proliferating economic division of labour and the emergence of internal social stratification and governance are emerging in the youthful mining settlements throughout the country's mining regions. The overall demographic picture is extremely dynamic. This contribution represents a preliminary step in documenting Tanzania's mining and urbanising synergetic trajectory. The 2012 census will provide a vital update for a decade in which Tanzania's mining has become the nation's leading economic sector. Identifying associated demographic trends is key to the formulation of timely spatial and economic policies. Furthermore, understanding the specific dynamics of urbanisation is requisite for prioritising infrastructural investment and social services to encourage productive and welfare-enhancing benefits.

Tanzania is in an enviable position at present. From 2001, its trade balance has been in surplus given its rising gold exports (Roe and Essex 2009, 28). In 2010, the value of the country's exports superseded imports (Bank of Tanzania 2010). The exceptionally high price of gold in the current global recession and the mineral richness of the country combined to provide an important window of opportunity in Tanzania's economic history. Tanzania's mineral treasure-trove must be secured as a credit rather than a curse by moving rationally from the eureka discoveries of small-scale miners to judicious national investment in infrastructure and well-being for the national population. Distributive justice in terms of jobs and residential settlement is paramount.

Timely infrastructural planning and investment in small towns can avert the problems of infrastructural congestion and inadequacy currently so apparent in Dar es Salaam and the secondary cities. Rapid, unplanned urbanisation and lack of economic diversification, generating a host of troubles in the form of labour displacement and impoverishment for vast numbers of Tanzanians caught in the transition from rural agrarian to urban career pursuits, is avoidable with careful planning. Tanzania faces a literal and figurative golden investment opportunity that should not be squandered. Recent Tanzania mining legislation has taken small-scale miners' interests into account, but mining's impact on the country will largely depend on distributive follow-up settlement planning, infrastructural investment and employment policies in an increasingly urbanised country.

Notes

1. 'Small-scale mining' is labour-intensive and involves low capital investment, a low degree of mechanisation, basic equipment and a high degree of occupational risk.
2. Decline was connected with a dispute over royal succession and the rising importance of Mombasa and Malindi in trade with India (Moon 2005).
3. Before the Portuguese built Fort Jesus, Mombasa had already become the favoured port of call for Indian sea-faring merchants given that the strength and timing of the monsoonal winds were superior to those of Kilwa (personal communication with Prof. Abdul Sheriff, 5 March 2011).
4. The value of gold exports as a percentage of total exports was 6% in 1930–1931 and 11% in 1936.
5. Lemelle (1986, 332) perceives the number to be an underestimation, as many of the miners from the neighbouring countries were not counted. At the time the population of Tanganyika would have been close to 6 million. Assuming that half the population was economically active, less than 1% would have been engaged in mining.
6. One of the current authors, Crispin Kinabo, participated in the study, which has yet to be published.
7. This entailed: 1) facilitating forward and backward linkages of mining with other sectors of the Tanzanian economy including the expansion of mining company's procurement of local goods and services in place of imports; 2) giving preference to the employment of Tanzania's trained mining experts over expatriates; and 3) narrowing the gap between foreign profits from mining and revenues to the Tanzanian government.
8. Bryceson and Mwaipopo (2010) found that approximately 20% of migrants to the diamond and gold mine sites they surveyed were born in urban areas.
9. Muzzini and Lindeboom's (2008, 80) study examined census data from 1988 and 2002 focusing on: the 'cities' of Dar es Salaam, Mwanza and Mbeya; the 'municipalities' of Moshi, Tabora, Iringa, Dodoma, Songea, Mtwara, Sumbawanga, Singida, Kigoma, Bukoba, Musoma, Arusha, Morogoro, Shinyanga, Kinondoni, Temeke, Ilala; and the 'towns' of Kibaha, Babati, Korogwe and Lindi (2008, 79) as well as 77 regional townships.
10. This was the case in Mwanza, Shinyanga, Mara, Rukwa and Mbeya regions. Only Kagera and Tabora regions, where mining activities are less significant, recorded larger regional capital growth relative to regional town growth (Muzzini and Lindeboom 2008, 52).
11. The e-Geopolis urban data set provides population estimates of Tanzanian settlements of over 10,000 people by decade throughout the twentieth century based on published census data as well as other official records. See http://http://www.e-geopolis.eu.
12. Mwanza and Shinyanga regions include the newly designated region of Geita.
13. There are other economic stimulants to growth in the gold ring area, notably a boom in Lake Victoria fisheries, rice and cotton farming, and livestock keeping.

Notes on contributors

Deborah Fahy Bryceson is a Reader in the Geographical and Earth Sciences School at the University of Glasgow. She has published on the theme of African de-agrarianisation and livelihood diversification, including *Farewell to Farms: De-Agrarianization and Employment in Africa* (Ashgate 1997), *Disappearing Peasantries: Rural Labour in Africa, Asia and Latin America* (2000), *African Urban Economies* (2006), and *How Africa Works: Occupational Change, Identity and Morality in Africa* (2010). She currently coordinates the Urbanization and Poverty in Mining Africa (UPIMA) research programme. During the last five years her work has concentrated on documenting the world of East African artisanal mining. She can be contacted at: dfbryceson@bryceson.net

Jesper Bosse Jønsson is Research Fellow at the School of Geographical and Earth Sciences at the University of Glasgow. He coordinated the Centre for Sustainable Artisanal and Small-Scale Mining (SASMIN), located at the University of Copenhagen, before taking up his current position as international field coordinator in the UPIMA research programme. He has recently published a book about the world travels of a Danish gold miner at the turn of the twentieth century entitled *Guldgraveren: Danskeren der fandt Kong Salomons Miner* and has published numerous journal articles on Tanzanian artisanal mining. He can be contacted at: jesper.jonsson@glasgow.ac.uk

Crispin Kinabo is a Senior Lecturer at the Department of Geology, University of Dar es Salaam, Tanzania. He has more than 20 years of professional experience on small-scale mining with an interest in its socio-economic aspects, demography and technology development for artisanal and small-scale mining. Since 1998 he has coordinated the Small Scale Mining Project within the Southern African Network for Training and Research on the Environment (SANTREN) and is a steering committee member on Minerals, Mining and Sustainable Development (MSSD) for the SADC countries. He has published several book chapters on the socio-economic impact of artisanal mining in developing countries. He can be contacted at: kinabo_2003@yahoo.co.uk

Mike Shand (FRSGS, FBCartS) is a GIS cartographer and Honorary Research Fellow at the University of Glasgow involved in GIS mapping, research and teaching. His career encompasses numerous mapping projects, research and consultancies in Tanzania, Nigeria, South Africa, India, Philippines, Saudi Arabia, Iran and Bosnia Herzegovina. Currently he is a team member of the DfID/ESRC-funded UPIMA research in Tanzania, Ghana, Angola; the Medical Research Council-funded project on 'Designing large-scale rabies vaccination programs in Tanzania, Philippines, Bali, South Africa'; and a BBSRC/ESRC/NIH-funded project on 'The impact and social ecology of bacterial zoonoses in northern Tanzania'. His recent publication is: *The wild heart of Africa, 110 years of the Selous Game Reserve, Tanzania* (2009). He can be contacted at: mike.shand@glasgow.ac.uk

References

Bank of Tanzania. 2010. *Economic and operations reports.* Dar es Salaam: Bank of Tanzania.
Beall, J., N. Kanji, and C. Tacoli. 1999. African urban livelihoods: Straddling the rural–urban divide. In *Urban poverty in Africa: From understanding to alleviation*, ed. S. Jones and N. Nelson, 160–68. London: Intermediate Technology Publications.
Bryceson, D.F. 1984. *Urbanisation and agrarian development in Tanzania with special reference to secondary cities.* London: International Institute for Environment and Development.
Bryceson, D.F. 1990. *Food insecurity and the social division of labour, 1919–1985.* London: Macmillan.
Bryceson, D.F. 1999. African rural labour, income diversification and livelihood approaches: A long-term development perspective. *Review of African Political Economy*, no. 80: 171–89.

Bryceson, D.F. 2006. Vulnerability and viability of East and Southern Africa's apex cities. In *African urban economies: Viability, vitality or vitiation?*, ed. D.F. Bryceson and D. Potts, 319–40. London: Palgrave Macmillan.

Bryceson, D.F. 2009. Sub-Saharan Africa's vanishing peasantries and the specter of a global food crisis. *Monthly Review*, July–August: 48–62.

Bryceson, D.F. 2011. Birth of a market town in Tanzania: Towards narrative studies of urban Africa. *Journal of Eastern African Studies* 5, no. 2: 274–93.

Bryceson, D.F., and J.B. Jønsson. 2010. Gold digging careers in rural Africa: Small-scale miners' livelihood choices. *World Development* 38, no. 3: 379–82.

Bryceson, D.F., and R. Mwaipopo. 2010. Rural–urban transitions in Tanzania's northwest mining frontier. In *Rural–urban dynamics: Livelihoods, mobility and markets in African and Asian frontiers*, ed. J. Agergaard, N. Fold, and K.V. Gough, 158–74. London: Routledge.

Bryceson, D.F., J.B. Jønsson, and H. Verbrugge. 2013 Forthcoming. Prostitution or partnership? Wifestyles in Tanzanian gold-mining settlements. *Journal of Modern African Studies* 51, no. 1.

Bryceson, D.F., J.B. Jønsson, and H. Verbrugge. Forthcoming. Loosely-woven love: Conjugal bonds and family formation in artisanal mining settlements. In *Mineralizing Africa: Artisanal mining and social change*, ed. D.F. Bryceson, E. Fisher, J.B. Jønsson and R. Mwaipopo. forthcoming.

Carstens, J., and G. Hilson. 2009. Mining and grievance in rural Tanzania. *International Development Planning Review* 31, no. 3: 301–26.

Chachage, C.S.L. 1995. The meek shall inherit the earth but not the mining rights: The mining industry and accumulation in Tanzania. In *Liberalised development in Tanzania*, ed. P. Gibbon, 37–108. Uppsala: Nordiska Afrikainstitutet.

Dickson, J. 1925. *Lupa diary.* Oxford University: Rhodes House, RH Mss. Afr. s.738.

Drechsler, B. 2001. *Small-scale mining and sustainable development within the SADC region.* London: International Institute of Environment and Development/World Business Council for Sustainable Development.

Emel, J., and M.T. Huber. 2008. A risky business: Mining, rent and the neoliberalisation of 'risk'. *Geoforum* 39, no. 3: 1393–407.

Geyer, H.S. 1996. Expanding the theoretical foundation of the concept of differential urbanisation. *Tijdschrift voor Economische en Sociale Geografie* 87, no. 1: 44–59.

Geyer, H.S., and T. Kontuly. 1993. A theoretical foundation for the concept of differential urbanisation. *International Regional Science Review* 17, no. 2: 157–77.

Hirst, M.A. 1972. A functional analysis of towns in Tanzania. *Tijdschrift voor Economische en Sociale Geografie* 64, no. 1: 39–51.

Ibn Battuta. 1962. *The travels of Ibn Battuta.* Trans. H.A.R. Gibb. 4 vols. London: Hakluyt Society.

Jønsson, J.B., and D.F. Bryceson. 2009. Rushing for gold: Mobility and small-scale mining in East Africa. *Development and Change* 40, no. 2: 249–79.

Jønsson, J.B., and N. Fold. 2009. Handling uncertainty: Policy and organisational practices in Tanzania's small-scale gold mining sector. *Natural Resources Forum* 33, no. 3: 211–20.

Jønsson, J.B., and N. Fold. 2011. Mining 'from below': Taking Africa's artisanal miners seriously. *Geography Compass* 5, no. 7: 479–93.

Kulindwa, K., O. Mashinadano, F. Shechambo, and H. Sosovele. 2003. *Mining for sustainable development in Tanzania.* Dar es Salaam: Dar es Salaam University Press.

Lange, S. 2006. *Benefit streams from mining in Tanzania: Case studies from Geita and Mererani.* Bergen: Christian Michelson Institute, R2006.1.

Lemelle, S.J. 1986. Capital, state and labor: A history of the gold mining industry in colonial Tanganyika 1890–1942. PhD diss., University of California at Los Angeles.

Mitchell, P. 2005. *African connections: Archaeological perspectives on Africa and the wider world.* Walnut Creek, CA: Altamira Press.

Moon, K. 2005. *Kilwa Kisiwani: Ancient port city of the East African coast.* Dar es Salaam: Tanzania Printers/Government of Tanzania.

Muzzini, E., and W. Lindeboom. 2008. The urban transition in Tanzania: Building the empirical base for policy dialogue, World Bank stock no. 37274. http://siteresources. worldbank.org/CMUDLP/ Resources/ Tanzania_wp.pdf.

Mwaipopo, R., W. Mutagwaba, W.D. Nyanga, and E. Fisher. 2004. *Increasing the contribution of artisanal and small-scale mining to poverty reduction in Tanzania*. London: Department for International Development.

Nyerere, J.K. 1972. *Decentralisation*. Dar es Salaam: Government Printer.

Pearson, M.N. 1998. *Port cities and intruders: The Swahili coast, India, and Portugal in the early modern era*. Baltimore: The Johns Hopkins University Press.

Phillips, L.C., H. Semboja, G.P. Shukla, R. Sezinga, W. Mutagwaba, B. Mchwampaka, G. Wanga, G. Kahyarara, and P.C. Keller. 2001. *Tanzania's precious mineral boom: Issues in mining and marketing*. Washington, DC: USAID.

Potts, D. 2010. *Circular migration in Zimbabwe and contemporary Sub-Saharan Africa*. Woodbridge, Suffolk, UK: James Currey.

Roberts, A.D. 1986. The gold boom of the 1930s in Eastern Africa. *Journal of the Royal African Society* 85, no. 341: 545–62.

Roe, A.R., and M. Essex 2009. Mining in Tanzania – What future can we expect? Paper presented at the ICMM Workshop on Mining in Tanzania, Dar es Salaam, May 18, 2009 (revised June 28, 2009). Oxford, Oxford Policy Management.

Shand, M. 2011. *Tanzanian small-scale mining sites relative to urban growth, 1950–2005*. Glasgow: University of Glasgow, UPIMA project.

Shand, M., and J.B. Jønsson. 2011. *Tanzania gold mines and minerals map*. Glasgow: University of Glasgow, UPIMA project.

Tan Discovery. 1996. *Final report on baseline survey and preparation of development strategy for small scale and artisanal mining program*. Dar es Salaam: Ministry of Energy and Minerals and Washington DC: World Bank.

Tanganyika Territory. 1938. *Report of the committee appointed to consider and advise on questions relating to the supply and welfare of native labour in the Tanganyika Territory*. Dar es Salaam: Government Printer.

Tanganyika. 1963. *African Census Report 1957*. Dar es Salaam: Government Printers.

Tanzania, United Republic of. 1970. *1967 population census*. Dar es Salaam: National Bureau of Statistics.

Tanzania, United Republic of. 1982. *1978 population census*. Dar es Salaam: Population Division, Bureau of Statistics.

Tanzania, United Republic of. 1991. *1988 population census*. Dar es Salaam: National Bureau of Statistics.

Tanzania, United Republic of. 1997a. *Mineral policy of Tanzania*. Dar es Salaam: Ministry of Energy and Minerals, Government Printer.

Tanzania, United Republic of. 1997b. *Investment Act of Tanzania*. Dar es Salaam: Government Printer.

Tanzania, United Republic of. 1998. *The Mining Act*. Dar es Salaam: Ministry of Energy and Minerals, Government Printer.

Tanzania, United Republic of. 1999. *The Mining Regulations*. Ministry of Energy and Minerals, Dar es Salaam: Government Printer.

Tanzania, United Republic of. 2005. *National strategy for growth and reduction of poverty*. Dar es Salaam: Vice President's Office, Government Printer.

Tanzania, United Republic of. 2008. *Bomani report of the Presidential Mining Review Committee to advise the government on oversight of the mining sector*. Vol. 2. Trans. Paperworks Associates Tanzania on behalf of Policy Forum. http://coet.udsm.ac.tz/biofuel%20documents/7.%20Bomani%20Report%20-20English.doc (accessed June 5, 2010).

Tanzania, United Republic of. 2009. *The new mineral policy of Tanzania*. Ministry of Energy and Minerals. Dar es Salaam: Government Printers.

Tanzania, United Republic of. 2010. *The Mining Act 2010*. Ministry of Energy and Minerals. Dar es Salaam: Government Printers.

Tanzania, United Republic of. 2012. www.nbs.tz/tnada/index.php.catalog/7. (accessed September 13, 2012).

Thompson, M. 1936–65. Melvyn's letters: A collection of letters and images written by and about Melvyn and Winona Thompson (Canadian geologist working in Tanzanian gold mining). http://www.melvynsletters.ca/.

Todaro, M. 1969. A model of labor migration and urban unemployment in less developed countries. *American Economic Review* 104: 387–413.
Todaro, M. 1997. *Urbanization, unemployment and migration in Africa: Theory and Policy.* New York: Population Council.
United Nations 2004. *World urbanisation prospects: The 2003 revisions.* New York: Economic and Social Affairs, Population Division.

Exploring the connections: mining and urbanisation in Ghana

Katherine V. Gough and Paul W.K. Yankson

Studies of mining and urbanisation have been primarily conducted independently of one another, with limited consideration of the inter-linkages between the two. This analysis seeks to fill this gap by exploring the links between mining and urbanisation in a Ghanaian context. Ghana is an interesting case as it is both endowed with significant mineral wealth and is highly urbanised, with a long history of urban settlement compared to most of sub-Saharan Africa. Mining and urbanisation are examined in four historical phases: pre-colonial; colonial; independence to the early 1980s; and from the early 1980s onward, before exploring the growth trends and employment characteristics of the large-scale and small-scale mining sectors, following which different types of urban mining settlements are discussed. The article shows that the fortunes of mining and the growth of urban settlements are interconnected but concludes that the relationship between the two is not one of simple cause and effect. The complex nature of the linkages between mining and urbanisation means that they require more attention than they have been afforded hitherto.

Much has been written about Ghana's mineral wealth since it was named the 'Gold Coast' by colonial powers in the nineteenth century. Similarly, there has been great interest in the pattern and nature of urban settlement in Ghana. These studies, however, have been primarily conducted independently of one another by mining and urban scholars respectively, with limited consideration of the inter-linkages between the two. Consequently, little has been written about how mining impacts on urbanisation or on the nature of urban mining settlements.

The intention here is to address this gap by exploring the links between mining and urbanisation in a Ghanaian context. Ghana is an interesting case as it is endowed with significant mineral wealth and has a long history of urban settlement. Minerals including gold, manganese, diamond, bauxite, limestone, silica and salt are all being exploited in commercial quantities, with gold representing by far the most important mineral mined. In addition to these, considerable resources of iron and various other industrial minerals exist (Akabzaa 2009). The history of mining in Ghana, however, is far from smooth and the production of gold has increased and decreased in line with changing government policy as well as prices on the world market. Within sub-Saharan Africa, Ghana is one of the most urbanised countries with an estimated 51% of the population living in urban areas[1] (Government of Ghana 2010).

In order to explore the connections between mining and urbanisation in Ghana, this contribution builds on secondary data, including raw data from the 2000 census, plus the authors' personal experiences of mining and urban settlements in Ghana. It is divided into two main sections. First, the links between mining and urbanisation are traced, dividing the analysis into four key phases: pre-colonial; colonial; independence to early 1980s; and early 1980s to today. Second, we turn to focus on the mining settlements themselves, examining their growth trends, employment characteristics and differing types of settlement. It is concluded that the links between mining and urbanisation are many but complex and deserve more attention than they have been afforded hitherto.

Mining and urbanisation trends in Ghana: key phases of development

Pre-colonial period

Mining has been carried out in the current territory of Ghana for centuries (Ofusu-Mensah 2011). Gold mining has an especially long history, with artisanal gold mining being one of the mainstays of numerous Akan states (Dumett 1979, 1998). Gold is known to have been one of the main items traded by the kingdoms, such as the Ashanti kingdom, from the Middle Ages. This gold was traded with North Africa, eventually reaching Europe along trans-Saharan trade routes. European traders, especially the Portuguese, are known to have traded in gold dust at the mouth of the Pra River from 1471 onwards (Allen 1958). At this time, all the gold being traded was won from stream and beach sands; hence it was not until the early eighteenth century that auriferous lodes as well as alluvial deposits were worked.

Pre-colonial African urbanisation is generally attributed to the rise of agriculture, regional and long distance trade, and powerful religious and administrative centres created by traditional leaders (Coquery-Vidrocitch 1991). During the pre-colonial period, the trans-Saharan trade routes resulted in towns such as Salaga and Bono Manso emerging as important trading centres in the north of present-day Ghana (Anderson and Rathbone 2000). Trading was associated with the development of farming, livestock rearing, fishing, salt-making, handicrafts and gold mining. As well as becoming centres of trade, political power and administration, these towns were also central to military operations in times of war (Kea 1982). Several towns linked to the Akan kingdoms and chieftaincies grew up close to mining centres (Dumett 1998; Kea 1982). Gold is thought to have been central to the establishment of the Great Ashanti Kingdom around 1700 (Ofusu-Mensah 2011). Kumasi, the base of the Ashanti kingdom, became the largest urban centre, being an administrative and spiritual centre as well as an important commercial centre based on technologies including iron smelting, stone building and gold mining (Anderson and Rathbone 2000). Trade with Europe also led to the emergence of trading towns along the coast, although Accra, which was later to become Ghana's capital, was at this time an indigenous settlement based on fishing. The continuing evidence of Ghana's indigenous urban imprint today is attributed to the country's fairly limited subsequent colonial European settlement (Otiso and Owusu 2008). As minerals were one of the products being traded in pre-colonial times, they contributed to the establishment and wealth of urban centres, but mining at the time did not directly result in urban settlement.

Colonial period

Modern gold mining that plumbs the rich ore deposits below the earth's surface began around 1860 in Ghana when European concessionaires imported heavy machinery and began working in the western areas of present-day Ghana. The turn of the century saw the so-called 'Jungle Boom' of 1901–1902 during which over 400 mining companies were formed. Although the majority of these companies never carried out mining operations, the large-scale equipping of mines took place and intensive prospecting and exploration of the gold areas commenced (Allen 1958). Gold mining was also boosted at this time by the extension of the railway line from the port of Sekondi, which reached Tarkwa in 1901, Obuasi in 1903 and Prestea in 1910 (Dickson 1969). Consequently the amount of gold being mined increased dramatically; official records indicate that the total amount of gold exported from the Gold Coast between 1471 and 1880 was 14.5 million ounces whereas from 1880 to 1955 total gold production was 22.3 million ounces (Allen 1958). Small-scale artisanal mining continued during the colonial period, employing thousands of people but featuring largely rudimentary and uncontrolled practices. Large-scale mining, on the other hand, involved significant capital investment from European investors. Under British rule, the colonial government controlled gold mining to protect the profits of European companies and restricted the possession of gold as well as mercury, which is essential in recovering gold from the ore in which it is embedded (Anyemedu 1992). By 1955, however, the exhaustion of ore reserves resulted in several companies closing down, reducing the number of gold producers in operation to 10.

During the colonial period, urbanisation occurred particularly along the coast as the emphasis on maritime trade increased. Cape Coast thrived as the initial capital of the Gold Coast and was the centre of trade, first for slaves and subsequently for the export of mineral and agricultural resources. Urban settlements also grew rapidly in the southern forest zone (nowadays the Ashanti, Eastern, Western and Brong Ahafo regions), linked to the exploitation of minerals alongside the expansion of cocoa farming and other cash crops. The building of the railways also contributed to urbanisation by enhancing the movement of goods and people. The focus of the European colonisers on the exporting of goods via the sea resulted in the northern towns which had developed during pre-colonial times being overlooked and declining (Songsore 2003a, 2003b). Some northern towns, however, such as Wa and Tamale, which were administrative centres, grew during the colonial period. The colonial rulers moved the capital along the coast to Accra in 1877 (Grant and Yankson 2003). The overall impact of the colonial period on urbanisation was the division of the country into a resource-rich urbanised southern part, partly based on mining activities, and a resource-poor less urbanised northern part.

Independence to the mid-1980s

Following independence in 1957, foreign control of Ghana's mining sector was tempered by increased government involvement under Nkrumah's regime. At that time many mines began to hit poorer gold reefs. These two factors resulted in few investors being willing to invest to expand production into new reefs, despite the floating of the international gold price in the late 1960s. As the government failed to provide the necessary capital, neither of the two major gold mining enterprises – the State Gold Mining Corporation and Ashanti Gold Corporation (which were 40% controlled by the government) – managed to maintain their production levels during this period. Production of gold thus started to decline in the late 1960s in line with the general decline in the economy. While in the decade 1960–1969 Ghana produced 8 million ounces of gold, production fell to just under 6 million ounces between 1970 and 1979, and by 1980–1989 this had gone down to just over 3 million ounces (Songsore, Yankson, and Tsikata 1994). Foreign exchange shortages inhibited mine maintenance, new exploration and development investment. The overvalued currency and spiralling inflation exacerbated mining companies' problems, as did smuggling and the deteriorating infrastructure. Energy supplies failed to meet the industry's growing needs; foreign exchange shortages constrained oil imports, and domestically generated hydroelectricity was unable to make up the shortfall (World Bank 1984). By the mid-1980s, only four gold mines were operating in Ghana (Songsore, Yankson, and Tsikata 1994). During this period, small-scale mining activities continued but remained largely unregulated and without any support from governmental bodies.

Following independence, all the major economic activities continued to be broadly located within a triangle from Accra and Sekondi-Takoradi on the coast to Kumasi inland (Figure 1). There was substantial migration into this area as people moved from all over Ghana to work in agriculture and mining. Following the adoption of a growth pole strategy, in 1975, regional development corporations established a range of ventures which further contributed to the growth of already existing urban settlements such as Cape Coast, Koforidua, Tamale and Ho (Yaro et al. 2011). Otiso and Owusu (2008) categorise this period as a national urbanisation phase, as development in Ghana was relatively insulated from the global economy and largely influenced by local and national forces, especially the government's national development policies. As this was a period of decline in the mining industry, the impact of large-scale mining on urbanisation was relatively limited, though people continued to migrate to the southern forest zone to work in small-scale mining, which was still not legalised by the state.

Mid-1980s to present

Since Ghana signed up to the Structural Adjustment Programme (SAP) prescribed by the World Bank and the International Monetary Fund in 1983, the mining sector has witnessed a considerable investment boom. The streamlining of policies, rehabilitation of infrastructure and improvement in the institutional and legal environment has resulted in increased investment flows. Ghana now has 16 operating mines, six projects at mine development stage and over 150 local and foreign companies with exploration licences, mainly in the domain of gold (Akabzaa 2009). Furthermore, the sector has attracted a significant number of mining support companies such as catering, transport and security companies, explosive manufac-turers and mineral assay laboratories. The sector's increasing contribution to the national economy is illustrated in Table 1. Total annual mineral exports rose to US$995 million in 2005, which corresponds to around a third of gross foreign exchange earnings. Gold is the most important sub-sector, accounting for over 90% of the total value of mineral exports in 2005. This level of production has been made

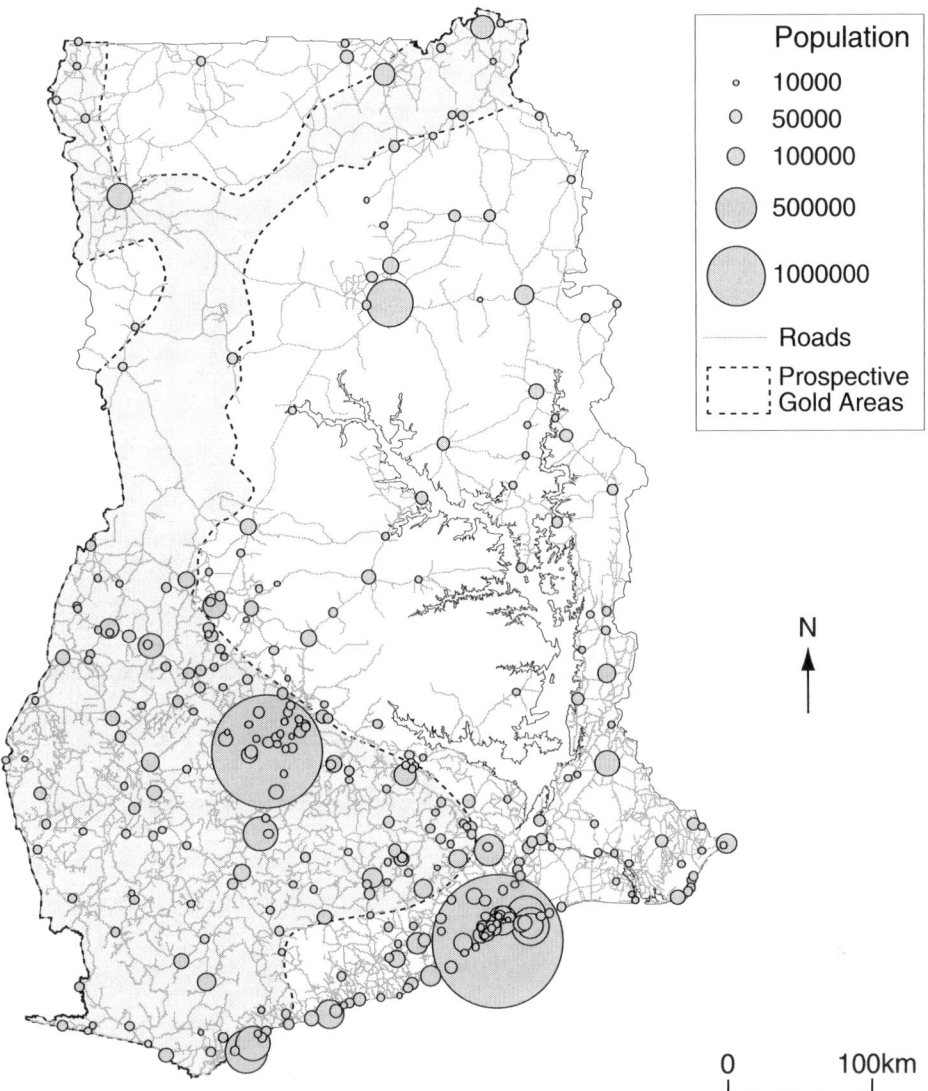

Figure 1. Location of largest urban localities and prospective gold areas in Ghana.
Source: based on Moller-Jensen and Knudsen (2008).

possible due to the increase in foreign direct investment (FDI) which rose to nearly US$6 billion by 2005, accounting for almost 60% of FDI flows to the Ghanaian economy (Akabzaa 2009).

From the mid-1980s onwards a number of laws were established to support small-scale mining (Songsore, Yankson, and Tsikata 1994). Not all small-scale operators, however, have registered with the Minerals Commission, although they are required by law to do so. Consequently, a large number of illegal operators, locally referred to as *galamseys*, engage in small-scale mining without licences and concessions. These miners extract gold from the alluvial deposits as well as from ore deposits spread across a number of regions including the traditional gold mining areas of the Western

Table 1. Contribution of the mining sector to gross exports value (1984–2005).

Exports	1984	1990	1995	2000	2005
Gold (US $)	103	202	647	702	946
Total mineral exports (US $)	115	242	679	756	995
Total exports (US $)	567	897	1431	1936	2836
Mineral as % of exports	20	27	47	39	35
Gold as % of exports	18	22	45	36	33
Gold as % of all minerals	90	83	95	93	95

Source: derived from Minerals Commission (2007) data in Akabzaa (2009).

and Ashanti Regions. New gold mining areas in the Brong Ahafo (Kenyease area) and Eastern Regions (Akim area), worked by the Newmont Mining Company, have also recently attracted a large number of *galamsey* operators. Parts of the Northern Region (Bole area) as well as the Upper East Region (Nagodi area) are also experiencing widespread small-scale mining activities. Throughout the adjustment period, however, the government has promoted the expansion of the large-scale sector to the detriment of the small-scale sector. The laissez-faire land concession allocation procedure, and the corresponding rapid rise in exploration and excavation activities by large-scale companies, has resulted in the displacement of thousands of subsistence artisanal gold miners. Although the Ghanaian government legalised small-scale mining in 1989, the complex and bureaucratic nature of the regulations and policies for *galamsey* miners favour the interests of large-scale mining. Despite these hindrances, the contribution of the small-scale sector to total gold production increased from US$3.4 million (2.2% of total gold production) in 1989 to US$79.8 million (9.5% of total gold production) in 2003 (Hilson and Potter 2005).

Although mining experienced a boom following the introduction of the SAPs, this was not the case for all sectors of the economy and there was a dramatic cut in the number of public sector employees. This resulted in widespread unemployment and previous government employees searching for new opportunities (Gough 2010). Due to the increased opening up of the economy and rising investment by foreign companies, Otiso and Owusu (2008) refer to this period as the 'global phase'. Investment has been greatest in the two major urban centres of Accra and Kumasi, resulting in their continuing dominance of the urban scene. Together these two cities account for just over a third (36%) of the total urban population and both are substantially larger than any other city (Table 2). Consequently, the level of urbanisation is non-uniform across the country, with the three northern regions (Upper East, Upper West and Northern Regions) substantially less urbanised, and poorer, than the southern regions (Figure 1). Compared to many other African countries, however, there are several secondary cities of a substantial size and Ghana has a lower level of urban primacy than many comparable countries. Interestingly, as Moller-Jensen and Knudsen (2008) report, many of the fastest growing urban settlements are relatively small and have only recently gained urban status; small urban settlements located close to high population concentrations have been growing especially quickly. Recent off-shore oil discovery in the western part of Ghana is also expected to result in a rapid increase in urbanisation in the area centred around Sekondi-Takoradi (Government of Ghana 2010).

Table 2. Ghana's 10 largest urban centres, 2000.

Settlement	Population
Accra metropolis	1,658,937
Kumasi metropolis	1,170,270
Takoradi sub-metropolis	289,593
Tamale metropolis	202,317
Ashiaman	150,312
Tema municipality	141,479
Obuasi	115,564
Sekondi sub-metropolis	114,157
Koforidua	87,315
Cape Coast	82,291

Source: Ghana Statistical Service (2002).

Mining continues to be a factor contributing to urbanisation in numerous ways. Obuasi, which is one of the largest 10 urban centres (Table 2), has grown predominantly based on mining activities. Numerous other smaller settlements have mining as their main economic activity. As will be shown below, however, the nature of urban settlements linked to mining varies greatly.

Changing levels of urbanisation and mining

Ghana's population has increased substantially since 1921 when the first formal census was undertaken. At this time the population was 2.3 million whereas by 2010 the population had risen more than 10-fold to 24.7 million (http://www.ghana.gov.gh).[2] This increase in the population has been mirrored by an increase in the proportion of the population living in urban areas (Table 3). From having a predominantly rural-based population, Ghana's urban population has increased, resulting in just over half of the population living in urban areas.

The cause of this urban growth is a combination of migration from rural areas to urban areas and a natural increase in the urban population. As Table 4 shows, the urban growth rate has been high (4.7/4.6%) in the intercensal periods 1960–1970 and 1984–2000, but lower (3.3%) in the period 1970–1984. Whereas in the 1960s urban

Table 3. Total population and level of urbanisation in Ghana.

Year	Total population (millions)	Percentage urbanised	No. of urban settlements
1921	2.30	7.8	–
1931	3.16	9.4	–
1948	4.12	12.9	41
1960	6.73	23.1	98
1970	8.56	28.9	135
1984	12.30	32.0	203
2000	18.91	43.8	364
2007	23.00	49.0	–
2009	23.80	51.0	–

Source: derived from Government of Ghana (2010) and Songsore (2003b).

Table 4. Contribution of migration and natural increase to urbanisation (%).

Intercensal year	Growth rate	Migration	Natural increase
1960–1970	4.7	54.5	45.5
1970–1984	3.3	25.0	75.0
1984–2000	4.6	37.4	62.6

Source: derived from Government of Ghana (2010).

growth was primarily due to migration, subsequently the majority of urban growth has been due to natural increase.

Whilst urbanisation has been a constant process, the fortunes of the mining industry have waxed and waned over the years. Overall mining clearly has been a factor contributing to urbanisation in Ghana but its precise impact is difficult to determine. As Figure 1 illustrates, there is a clear overlap between the areas with the majority of urban settlements and the prospective gold mining areas. It would be too simplistic to infer a direct correlation between the two, however, as the main mining areas are also the most fertile agricultural areas. Consequently, there has been considerable in-migration to work in agriculture, particularly on cocoa farms. As this account has shown, however, since pre-colonial times mining has been important to Ghana's economy and has contributed to the level of urbanisation and the location of some urban settlements. Having traced the linkages between the histories of mining and urbanisation in Ghana, we now turn to focus specifically on mining settlements, examining their growth trends, employment opportunities and differing settlement types.

Mining settlements in Ghana

Trends in growth and decline of mining settlements

Although gold mining has been carried out in Ghana for centuries, the establishment of the earliest gold mining settlements took place in the late nineteenth century (Dickson 1969). Population data to track the growth of mining settlements, however, is only readily accessible from 1960 onwards when the first widely available population census was conducted. Eight mining urban settlements have been identified by Yaro et al. (2011) as existing at this time. Obuasi, located in the Ashanti region, was clearly the largest with a population of just over 20,000, followed by five settlements with a population of 10–13,000, and two half that size, mostly located in the Western Region (Table 5).

During the period 1960 to 1970, three new urban mining settlements emerged – Awaso, Bibiani and Sefwi Bekwai – all located in the Western Region (Yaro et al. 2011). In all three cases the settlements grew within 10 years from a small rural settlement with a population of under 2000 to urban centres with populations ranging from approximately 13,000 to 35,000 (Table 6). This rapid population increase of over 1000% for the 10-year period was due to the discovery and exploitation of minerals in these settlements.

During the 14-year period between the next two censuses (1970–1984), a total of 25 settlements were identified as becoming urban settlements (i.e. gaining a population of over 5000) based on mining activities. The percentage increase in the

Table 5. Mining urban settlements in Ghana, 1960.

Settlement	Population
Obuasi	22,818
Tarkwa	13,545
Prestea	13,246
Dunkwa-on-Offin	12,689
Akwatia	12,592
Konongo	10,771
Odumase	5,540
Abosso	5,095

Source: Ghana Statistical Service (1989).

settlements during this period was not as high as during the previous period, however, due to the reduced investment in mining during this time (Yaro et al. 2011). The six settlements that experienced the highest rates of population increase are listed in Table 7. Most notable are Achiase in the Eastern Region and Dambai in the north of the Volta Region, which grew by around 800% and 500% respectively over this period. The only new urban settlement that is reported to have emerged due to mining activities during the subsequent inter-censal period (1984–2000) is Bogoso (Western Region), which increased from a population of 4536 to 8659.

Although over time new urban settlements based on mining activities have emerged, increasing the total number of mining settlements, the fortune of existing mining settlements has varied. The growth, and decline, of the eight mining settlements identified in 1960 are tracked in Table 8. Obuasi has clearly maintained its position as the most important mining settlement in Ghana, increasing in size more than five-fold in a period of 40 years and today being one of the largest settlements in Ghana. Most of the other settlements in Table 8 have more or less doubled in size during the same period. However, if the data are studied closely it can be seen that the story of the mining settlements is not just one of growth. In the period between 1960 and 1970, the populations of Prestea, Tarkwa, Akwatia, Odumase and Abosso all declined and, apart from Obuasi, only increased slightly in the other settlements. Prestea and Tarkwa are notable for their extreme decline in population. This is most likely due to the implementation of an Aliens Compliance Order by the Government of Ghana in 1969, which compelled aliens without a residence permit to leave the country. Prestea and Tarkwa were severely affected as they had many resident non-Ghanaians of West African origin. Between 1970 and 1984, Abosso was the only mining settlement to experience a drop in the population, although the population of Dunkwa only increased marginally. Obuasi, Tarkwa,

Table 6. Urban settlements that emerged due to mining, 1960–1970.

Settlement	1960	1970	Percentage increase
Awaso	548	35,449	6368.8
Bibiani	1942	29,691	1428.9
Sefwi Bekwai	947	13,118	1285.2

Source: derived from Yaro et al. (2011).

Table 7. Urban settlements that emerged due to mining, 1970–1984.

Settlement	1970	1984	Percentage increase
Achiase	1122	10164	805.9
Dambai	823	5210	533
A. Bremang	1560	5871	276.3
Gushiegu	2847	8945	214.2
Karaga	2638	6310	139.2
Zabzugu	2967	6846	130.7

Source: derived from Yaro et al. (2011).

Prestea and Odumase all experienced rapid population growth during this period. All of the listed settlements grew between 1984 and 2000, most notably Obuasi, Konongo and Abosso which approximately doubled in size.

Migration has been an important factor in the growth of all the mining towns. The mining settlements in the former Wassa West District of the Western Region serve as a useful example. The population of each of the urban localities in the district contains a large migrant segment, as indicated by place of birth (Table 9). In all the settlements between around a third and a half of the population are migrants, while in Bogoso a majority (53%) of the population are migrants.

The main gold mining towns which have active large-scale mining operations today are Obuasi in the Ashanti Region and Tarkwa, Prestea, Abosso, Damang, Bogoso and Bibiani, all in the Western Region. Konongo-Odumasi in the Ashanti Region had large-scale gold mining in the past but the mine is no longer operating. The same is the case for Dunkwa-on-Offin in the Central Region, which had a state company operating alluvial works in the past but the company is now on the divestiture list. Anyanfuri in the Central Region has a gold mining company operating and in Kenyiase, in the Brong Ahafo Region, the Newmont mining company has started gold production and is also prospecting in the Akim area of the Eastern Region. The key diamond-producing town is Akwatia, but the Ghana Consolidated Diamond Company has now stopped production and is on the divestiture list.

This survey of the changing population of mining settlements has shown how they can increase rapidly in size when a mineral is discovered but also decline rapidly if the mineral is exploited, which is symptomatic of mining towns elsewhere.

Table 8. Changing population of early mining urban settlements in Ghana.

Settlement	Population 1960	Population 1970	Population 1984	Population 2000
Obuasi	22,818	31,005	60,617	115,564
Tarkwa	13,545	4,702	22,107	30,631
Prestea	13,246	5,143	16,922	21,844
Dunkwa	12,689	15,437	16,905	26,215
Akwatia	12,592	12,177	15,007	20,723
Konongo	10,771	10,881	13,677	26,735
Odumase	5,540	4,624	8,779	13,903
Abosso	5,095	4,994	4,666	9,942

Source: 1960 data Yaro et al. (2011); other data Ghana Statistical Service (2002).

Table 9. Birthplace of Ghanaians in the urban localities of Wassa West District, 2000.

Settlement	Born in locality (%)	Born elsewhere in Western Region (%)	Born in other region in Ghana (%)
Tarkwa	52	15	33
Prestea	63	11	27
Abosso	62	15	23
Bogoso	47	20	33
Nsuaem	69	14	16

Source: 2000 Population and Housing Census of Ghana (Ghana Statistical Service 2002).

Consequently, a higher proportion of the inhabitants of mining settlements are migrants than in non-mining towns. In the next section we turn to the nature of employment in mining settlements in Ghana, which also directly impacts on changes in the population size of the settlements.

Changing levels of mining employment

The mining sector accounted for approximately 3% of the economically active population in the 1960, 1970 and 1984 censuses but this fell to 1.3% in the 2000 census. While large-scale mining provides direct employment for approximately 15,000 workers, it is estimated that about 300,000 people are engaged in small-scale mining (Hilson and Potter 2005). Thus the massive investment in mining, mainly by multinational companies since the early 1990s, has not resulted in any significant labour expansion in the formal mining sector. This is because of the application of more efficient, capital-intensive technology applied to surface mining, in contrast to the labour-intensive underground mining operations of the past. Mining is one of the major sectors in which the redeployment exercise associated with the SAPs has been most intense; from 1983 to 2003, over 15,000 workers from various mines lost their jobs. The retrenchment of a large number of unskilled and semi-skilled mine workers has created a serious unemployment and poverty situation, as seen in the mining settlements in the Wassa West District (Yankson 2010a). A substantial number of former workers of large-scale mining companies who have been laid off have become *galamseys*. These miners move from one area to another depending on the richness of the ore found in different locations, usually working without a legal land lease and very often on the concessions of large mining companies. Interestingly, while there has been a progressive reduction in the number employed in the large-scale mining sector, the quota of expatriate employees has kept on growing as a result of increased migration quotas for expatriate staffing in mining. The percentage of expatriate staff to Ghanaian senior staff increased from 8.8% to 11% between 1994 and 2006 (Akabzaa 2009).

The rapid rise in exploration and excavation activities by large-scale mining companies has resulted in the displacement of thousands of previously undisturbed small-scale gold miners (Akabzaa and Darimanni 2001; Hilson 2001, 2004; Hilson and Potter 2005). Land-use conflicts between small-scale miners and large-scale mining companies have escalated in recent years due, in large part, to decreased prospects for underground mining and the rapid rise in open-cast mining activity.

From the small-scale miners' perspective, the conflict is a contest for survival and a fight against the intrusion of what they perceive as their traditional rights to work on the land, whether for mining or farming (Aubynn 2009). As a result of conflicts with large-scale mining companies, many miners involved in illegal small-scale mining have had confrontations with the state security agencies resulting in some losing their employment. Conflicts have also arisen between the *galamsey* and local communities as the mining activities at times pose a serious threat to the surrounding environment; the miners rarely adhere to environmental and land-use regulations, occasionally spilling cyanide and employing crude mining methods that leave behind gaping holes which fill with water, contributing to the spread of malaria (Boateng 2003).

Although some towns may be considered to be mining settlements, the employment generating capacity of mining, especially large-scale, is rather limited. The case of the mining centres in the former Wassa West District of the Western Region, where between 15% and 26% of the economically active population is engaged in mining and quarrying, illustrates this situation (Table 10). In most of the settlements, wholesale/retail is the most important economic activity, whilst in Nsuaem agriculture employs the greatest number of people. The importance of mining has changed over time for many of these settlements. In Prestea, mining and quarrying employed 22% in 1960, which fell to 6% in 1984, but by 2000 this had increased to 26%. This was probably due to employment in the small-scale mining sector since none of the rehabilitated mines or the large- and medium-scale mining companies that were granted concessions following SAP were located in Prestea. On the other hand, although Tarkwa has grown in size, the occupational structure of the town has changed little since the 1960s. Mining and quarrying are the most important growth industries but only employed about 12% of the labour force (and virtually no women) in 1960 and 14% in 2000. Commerce is the most important sector for employment, engaging 30% of the working population, especially women. Services constitute a substantial proportion of the category 'other', and manufacturing provides 14% of employment (Table 10). Consequently, even at the height of the mining-related employment boom in these towns, their employment structure was not dominated by mining but by commerce and other non-farm activities (Yankson 2010a). We now turn to examine in more detail the nature of different types of mining settlements.

Types of mining settlement

The nature of settlements that have been established where mining is the predominant activity varies widely depending on a range of factors including the

Table 10. Employment by sector (%) in towns in Wassa West District, 2000.

Settlement	Agriculture	Mining and quarrying	Manu-facturing	Wholesale/retail	Other
Tarkwa	6	14	15	30	36
Prestea	13	26	19	23	22
Bogoso	17	17	16	27	23
Abooso	27	15	15	20	24
Nsuaem	43	18	11	11	16

Source: Compiled from raw data: 2000 Population and Housing Census of Ghana (Ghana Statistical Service 2002).

type of mining engaged in and the age of the settlement. Some of the key gold mining towns such as Obuasi and Prestea were created entirely by gold mining companies. These towns developed on mine concessions and they have grown and assumed other roles over time. Other towns developed in agriculture and other resource-rich areas and became mining centres when mineral explorations started within their vicinities – for example, the Tarkwa–Aboso–Nsuta–Bogoso mining complex in the Wassa West District in the Western Region. None of the mining towns can be described as specialised mining towns in terms of their employment structure. Here, this diversity is illustrated through presenting the profiles of three very different settlements: Obuasi, which is the largest and one of the oldest mining settlements and the base for some of the most important large-scale mining in Ghana; Tarkwa, which is the oldest mining settlement in an area which has both formal and informal deep-shaft mining; and Kui, which is a new mining settlement established in an area of informal opencast mining.

Obuasi

Gold mining was carried out in Obuasi for centuries by Akan groups and by the Ashanti from the start of the eighteenth century (Hilson 2002). Almost 200 years later, in 1895, the British opened a series of gold mines in Obuasi, the most important being operated by the Ashanti Goldfields Company. The arrival of a railway line from Sekondi in 1903, which continued on to Kumasi, was a further boost to settlement in Obuasi (Dickson 1969). Today Obuasi has the most important gold mine in Ghana and is also one of the world's richest mining operations (Hilson 2002).

Ever since an agreement was signed between the first European Company and the Chief of Bekwai, workers from all over Ghana and elsewhere in West Africa have migrated to Obuasi to make a living (Ampene 1965). The population of Obuasi increased rapidly from around 20,000 in 1960 to just over 115,000 in 2000, making it one of the largest towns in Ghana (Table 8). Mining has remained a key sector of the economy in Obuasi despite those directly employed in mining being a minority and having decreased over the years. The capital-intensive mechanisation programme implemented during the 1980s and 1990s resulted in staffing at Ashanti Goldfields Company being reduced from approximately 10,000 in 1996 to less than 6600 in 2003 (http://www.mining-technology.com/projects/obuasi). The town's economy is not only dependent on mining, but has diversified through commerce and cocoa production, which has attracted migrant labour from northern Ghana. Obuasi is also noted for its advanced health care facilities. Mining, however, continues to be the mainstay of the economy and the principal reason for the existence and growth of Obuasi.

Tarkwa

Tarkwa is the oldest centre of formal gold mining in Ghana. The initial settlement was established through gold mining in the 1880s, but Tarkwa's subsequent growth was also due to its nodal position at the centre of several trading routes (Dickson 1969). The arrival of the railway from the port of Sekondi in 1901 increased the role Tarkwa played as an important trading and administrative town in the region. Consequently, Tarkwa became a magnet for in-migration and the population increased rapidly from approximately 8000 in 1948 to over 13,500 in 1960. Subsequently, the population fell, in part due to the demise of the mining industry, but by 2005 the population of Tarkwa (together with the surrounding villages) was estimated by planning officials in the District Assembly to be around 53,000 (Yankson 2010a). In the early period, the population was ethnically mixed and consisted of Fanti, Appolonia and Ashanti tribesmen who gained a livelihood by working in the mines, acting as carriers and running liquor stores. When the mining went underground there was large-scale recruitment of labour from northern Ghana as the local Wassaw people had an aversion to working underground (Hug 1989, cited in Songsore, Yankson, and Tsikata 1994). Consequently, a significant proportion of the population in Tarkwa are second- or third-generation migrants (Songsore, Yankson, and Tsikata 1994).

Tarkwa continues to have an important administrative role as the capital of Wassa West District. Mining remains an important sector with seven large-scale foreign mining companies and a large number of unregistered small-scale mining groups. The recent influx of migrants to Tarkwa has resulted in the physical expansion of the town, though much of this has occurred without physical planning or the adequate expansion of infrastructure. There is severe overcrowding (Yankson 2010a) and some young migrants have taken to sleeping in the railway station and lorry park (Akabzaa and Darimani 2001). The boom in mining has stimulated the service and commerce sectors, and the availability of drugs is also reported to have increased (Akabzaa and Darimani 2001). Although employment generation by large-scale mining companies has been limited, the economy of the town is closely linked to the fortunes of the mining industry. When the mining economy suffers a downward trend the whole town economy suffers. Currently, however, with high gold prices the mining sector is thriving and Tarkwa is booming (Yankson 2010b).

Kui

Kui is a new mining settlement established about four years ago in the Bole district in the Northern Region. It is reached by travelling about 20 km from Tinga down a very rough dirt track that is only passable by four-wheel drive vehicles at any time and only by motorbikes in the rainy season. The settlement is based on the open-cast extraction of gold by *galamsey* operations. An estimated 4000–5000 people live in Kui all year round, though this number rises in the rainy season, which is the peak mining activity. People have migrated to Kui from all over Ghana and beyond in the hope of making their fortune. The numerous ethnic groups that are resident in Kui are reflected in the nature of the jobs that they undertake and the location of their residence. The Dagabas are dominant in underground mining in the dry season, a number of them having had experience of working previously in the mines in Obuasi and Tarkwa, while the local Gonja women predominate in food preparation and petty trading. None of the structures in Kui are permanent and almost all of them are built of basic materials covered with plastic sheeting in an attempt to keep out the rain. A few better-constructed wooden structures, however, are starting to appear (Kala and Gough 2010).

Given the difficulty of access to Kui, the residents rarely leave the settlement; hence, extensive service and commercial sectors have arisen to provide for the miners and their families. An impressive array of goods is on sale, far wider than is the norm for a settlement of this size, reflecting the wealth that can be generated by mining. Infrastructure provision, however, is very limited, with none having been provided by the District Assembly as Kui is regarded as an illegal settlement. A multitude of generators, owned by enterprising individuals, provide electricity and a number of private individuals have also dug boreholes to sell water. As there are no sanitation facilities, the surrounding bush is being used (Kala and Gough 2010).

There is a strong likelihood that once the gold has been exhausted the miners will up camp and move elsewhere and Kui will cease to exist. The impact on nearby Tinga, however, may be longer lasting and a number of new houses constructed in Tinga have already been attributed to the gains from mining activities in Kui (Kala and Gough 2010). This raises the important issue of where miners invest their gains from mining activities. As many of those engaged in mining are migrants, they often invest in urban settlements located elsewhere. In Kui, for example, a miner recounted how with the money gained from mining he had been able to construct four houses in different locations – Tinga, Wa, Techiman and his hometown Sankara – as well as establish two drinking spots (bars) in Kui. This highlights how the impact of mining on urbanisation extends well beyond the mining settlements themselves.

Conclusion

This paper has shown how the fortunes of mining and the rise of urban settlements in Ghana are interconnected but that the relationship between the two is far from clear-cut. Mining is clearly part of the story of urbanisation in Ghana but it is far from the whole story. Mining, and the corresponding trade in minerals, were the reasons behind the establishment of some of the early settlements in Ghana. Over time, the location of the majority of urban settlements has been shown to coincide with the areas in which gold is (or has been) mined. Some urban settlements in these areas have grown due to the in-migration of miners and their families. The area is also agriculturally rich, however, and there has also been substantial migration and urban growth due to employment possibilities in the agricultural sector, consequently a simple cause and effect between mining and urban settlement cannot be established. Furthermore, the transient nature of many miners means that often they are investing, and contributing to urbanisation, in locations other than where the mines are located. This is particularly the case with operators in small-scale, often illegal mining activities, whereas many miners working in the formal mining sector have settled and invested in the mining towns themselves.

Only one of the top 10 urban centres in Ghana, Obuasi, has mining as its economic base. Many mining settlements remain relatively small urban centres, their size varying in accordance with the fortunes of the mineral upon which they are based. Although mining only provides direct employment for a minority of inhabitants in mining urban settlements, widespread employment is generated in a range of subsidiary and supporting activities. Some mining settlements are primarily transitory settlements and cease to exist once the mineral has been exploited, whereas others take on additional roles and continue to thrive even if mining activities decline.

Despite the importance of mining in Ghana's economy, it is rarely mentioned in accounts of urbanisation in Ghana except in passing. The recently written Draft National Urban Policy (Government of Ghana 2010), for example, scarcely mentions mining. The growth pole strategy for urban settlement which is developed in the document does not appear to take mining into account. Given the importance of mining in the economy, and the number of settlements which are established due to mining, this is rather surprising. Furthermore, the importance of mining in Ghana's economy means that the fortunes of the mining sector directly affect the strength of the Ghanaian economy, which in turn impacts on urbanisation in general. This paper has just started to examine these links. Further studies, in Ghana and elsewhere, that explore the links between mining and urbanisation in more detail are needed and could in due course make a valuable contribution to policy making in sub-Saharan Africa and beyond.

Acknowledgements

The authors would like to thank the special issue editors, Deborah Bryceson and Danny MacKinnon, as well as two anonymous referees for their helpful comments on this paper. They would like to thank Lasse Moller-Jensen for giving permission to use his map of Ghana and Mark Szegner for editing the map.

Notes

1. It is important to note that settlements with a population of 5000 or more are recorded as being urban in Ghana.
2. At the time of writing the total population of Ghana from the 2010 census had been released but the population of individual towns was still unavailable.

Notes on contributors

Katherine V. Gough is a Reader in Department of Geography, Loughborough University and Adjunct Associate Professor in the Department of Geography and Geology, University of Copenhagen. She researches on urban issues in the global South, contributing to debates concerning: housing and home; youth, mobility and entrepreneurship; rural–urban dynamics; urban governance and planning; and comparative urbanism. She has conducted research in Latin America, Africa and the Asia-Pacific region and has a particular interest in comparative and longitudinal research. Recent books include *Rural–urban dynamics: Livelihoods, mobility and markets in African and Asian frontiers* and *Youth and the city in the global south: Urban lives in Brazil, Vietnam and Zambia*. She can be contacted at: K.V.Gough@lboro.ac.uk.

Paul W.K. Yankson is Professor of Geography at the University of Ghana, Legon. His current research projects include: urbanisation and poverty in mining Africa; youth and employment: entrepreneurship in African economies; and small-scale gold mining in Ghana: the gold chain. His latest publication in the area of mining is: Gold mining and corporate social responsibility in the Wassa West District, Ghana. 2010. *Development-in-Practice* 20, no. 3: 354–66. He can be contacted at: pyankson@ug.edu.gh.

References

Akabzaa, T. 2009. Mining in Ghana: Implications for national economic development and poverty reduction. In *Mining in Africa Regulation and Development*, ed. B. Campbell, 25–65. New York: Pluto Press.

Akabzaa, T., and A. Darimani. 2001. *A Study of impacts of mining sector investment in Ghana on mining communities*. Accra: Structural Adjustment Participatory Review Initiative (SAPRI) on Ghana. http://www.naturalresources.org/minerals/africa/docs/pdfs/Impact%20of%20Mining%20Sector%20Investment%20in%20Ghana.pdf (accessed March 26, 2009).

Allen, G.T. 1958. Gold mining in Ghana. *African Affairs* 57: 221–40.

Ampene, E. 1965. Individual research report: A study of urbanization – Progress report on Obuasi project. http://archive.lib.msu.edu/DMC/African%20Journals/pdfs/Institue%20of%20African%20Studies%20Research%20Review/1966v3n1/asrv003001005.pdf (accessed December 27, 2008).

Anderson, D.M., and R. Rathbone. 2000. Urban Africa: Histories in the making. In *Africa's urban past*, ed. D.M. Anderson and R. Rathbone, 1–17. Oxford: James Curry.

Anyemedu, K. 1992. Planning for a wider impact of the mining industry on the community. In *Planning African growth and development – Some current issues*, ed. K. Aryeetey, 322–47. Accra: ISSER/UNDP.

Aubynn, A. 2009. Sustainable solution or a marriage of inconvenience? The co-existence of large-scale mining and artisanal and small-scale mining in the Abosso goldfields concession in Western Ghana. *Resource Policy* 34, nos. 1–2: 64–70.

Boateng, J.Y. 2003. *Enhancing business–community relations: Ghana mining industry-case study.* http://www.new-academy.ac.uk/www.agi.org.gh/www.unv.org.

Coquery-Vidrovitch, C. 1991. The process of urbanization in Africa (from the origins to the beginning of independence). *African Studies Review* 34, no. 1: 1–98.

Dickson, K.B. 1969. *A historical geography of Ghana.* Cambridge: Cambridge University Press.

Dumett, R.E. 1979. Precolonial gold mining and the state in the Akan region: With a critique of the Terray hypothesis. In *Research in economic anthropology*, ed. G. Dalton, 37–68. Greenwich, CT: JAI Press, Inc.

Dumett, R.E. 1998. *Eldorado in West Africa: The gold-mining frontier, African labor, and colonial capitalism in the Gold Coast, 1875–1900.* Oxford: James Currey.

Ghana Statistical Service (GSS). 2002. *2000 Population and Housing Census: Special Report on 20 Largest Localities.* Ghana: GSS.

Gough, K.V. 2010. Continuity and adaptability of home-based enterprises: A longitudinal study from Accra, Ghana. *International Development Planning Review* 32, no. 1: 45–70.

Government of Ghana. 2010. *Draft national urban policy background document.* Accra.

Grant, R., and P.W.K. Yankson. 2003. City profile: Accra. *Cities* 20, no. 1: 65–74.

Hilson, G. 2001. A contextual review of the Ghanaian small-scale industry. *Mining Minerals and Sustainable Development* 76: 1–29.

Hilson, G. 2002. Harvesting mineral riches: 1000 years of gold mining in Ghana. *Resources Policy* 28: 13–26.

Hilson, G. 2004. Structural adjustment in Ghana: Assessing the impact of mining sector reform. *Africa Today* 51, no. 2: 53–77.

Hilson, G., and C. Potter. 2005. Structural adjustment and subsistence industry: Artisanal gold mining in Ghana. *Development and Change* 36, no. 1: 103–31.

Kala, M., and K.V. Gough. 2010. Galamsey operation and poverty reduction: Changing livelihoods of young miners in Northern Ghana. Paper presented at the African Studies Association Annual Conference, Oxford, September 17.

Kea, R.A. 1982. *Settlement, trade and politics in the seventeenth-century Gold Coast.* Baltimore and London: Johns Hopkins University Press.

Moller-Jensen, L., and M.H. Knudsen. 2008. Patterns of population change in Ghana (1984–2000): Urbanization and frontier development. *Geojournal* 73: 307–20.

Ofosu-Mensah, E.A. 2011. Historical overview of traditional and modern gold mining in Ghana. *International Research Journal of Library, Information and Archival Studies* 1, no. 1 (August): 6–22. http://www.interesjournals.org/IRJLIAS (accessed May 21, 2012),

Otiso, K.M., and G. Owusu. 2008. Comparative urbanization in Ghana and Kenya in time and space. *Geojournal* 71: 143–57.

Songsore, J. 2003a. *Regional development in Ghana: The theory and the reality.* Accra: Woeli Publishing Services.

Songsore, J. 2003b. *Towards a better understanding of urban change: Urbanization, national development and unequality in Ghana.* Accra: Ghana Universities Press.

Songsore, J., P.W.K. Yankson, and G.K. Tsikata. 1994. *Mining and the environment: Towards a win–win strategy. A study of the Tarkwa–Aboso–Nsuta mining complex.* Accra: Final report prepared for the Ministry of Environment, Science and Technology (MEST-EPC)

and the World Bank in support of the formulation of a growth strategy for the Western Region.

World Bank. 1984. *Ghana: Policies and program for adjustment*. Washington, DC: World Bank Publishing.

Yankson, P.W.K. 2010a. The dynamics of the gold mining industry and its effects on settlements and livelihoods in Wassa West District, Ghana. In *Rural–urban dynamics: Livelihoods, mobility and markets in African and Asian frontiers*, ed. J. Agergaard, N. Fold, and K. Gough, 175–88. London and New York: Routledge.

Yankson, P.W.K. 2010b. Gold mining and corporate social responsibility in the Wassa West District, Ghana. *Development in Practice* 20, no. 3: 354–66.

Yaro, J.A., S.N.A. Codjoe, S. Agyei-Mensah, A. Darkwah, and S.O. Kwankye. 2011. Migration and population dynamics: Changing community formations in Ghana. Center for Migration Studies, University of Ghana, Migrations Studies Technical Paper Series, Technical Paper no. 2.

A tale of two cities: urban transformation in gold-centred Butembo and diamond-rich Mbuji-Mayi, Democratic Republic of the Congo

Patience Kabamba

The recent wars in the DRC are partly caused and prolonged by the existence of lootable resources and have precipitated one of the most severe humanitarian catastrophes of our era in terms of loss of lives and human misery. Nonetheless, certain groups have taken advantage of both the absence of the state and the presence of minerals to modernise and institute new forms of order and development. This contribution provides an anthropological interpretation of the differences in the urbanisation of two mineral cities: Butembo and Mbuji-Mayi, located in eastern and central DRC respectively. Butembo is built on gold mines while Mbuji-Mayi sits on the biggest concentration of industrial diamonds in central Africa. After a critical survey of different approaches to the pace of urbanisation in mineral-rich DRC, an alternative explanation of the differences between the two cities is presented. This rests on the presupposition that processes of urbanisation are shaped by cultural legacies, which make the social pressure for redistribution stronger in some secondary cities than others.

For the last two decades the Democratic Republic of the Congo (DRC) has in the minds of many resembled a beggar with diamonds in his pocket. Congo is comparable to the European Union in size and extremely well endowed with minerals. Commentators are often ecstatic about its natural resources:

> The mineral belt that fans out from Katanga's dry savannah into neighbouring Zambia contains copper and zinc in concentrations rival nations can only dream about and enough cobalt to corner the global market. Nearly 500 miles North-West lies another gift of nature: the dark red gravel banks that trace the winding course of the Kanshi River, second biggest source of industrial diamonds in the world. There are other gifts: diamonds at Tshikapa in the South-West and Kisangani in the North, what was for a time the world's main source of uranium at Shinkolobwe, and across from the border with Uganda comes the enticing glitter of gold, cadmium and cassiterite, manganese and wolframite, beryl, columbo-tantalite and germanium: metals with mysterious, evocative names. No wonder a US ambassador once memorably referred to 'the Congo caviar' in a cable back to headquarters. (Wrong 2000, 108)

Yet the Congo has recently joined the group of extremely poor and deeply indebted countries and is currently ranked at 187 out 187 countries in the United Nations' human development index (UNDP 2011). The contrast between its known wealth and the economic welfare of its population has prompted people to call the Congo the most tragic example of a natural resource curse. Far from progressing on a path of speedy modernisation, Congo's natural resources are a drawback, attracting a motley assortment of foreign investors looking for large profits and Congolese compradors happy to benefit personally at the expense of national wealth distribution. The current protracted state of civil war in the DRC is partly caused, and undoubtedly fuelled, by the existence of the vast reserve of lootable mineral resources. The result has been an unprecedented humanitarian catastrophe of greed. The Congolese conflict has caused unparalleled levels of suffering. By 2006, 4 million people had died out of a population of 58 million. Seven million suffered from malnutrition, 3 million were HIV positive, 2.4 million were internally displaced, 880,000 had become refugees, 3 million children were orphaned and at least 40,000 women had been victims of sexual violence.[1]

Nonetheless, while the majority of people and cities suffer, certain groups have been able to take advantage of both the absence of the nation-state and the presence of minerals to modernise and institute new forms of order and development. Mbuji-Mayi, a city built on the most important diamond site in Central Africa, has suffered enormously from the destruction of the war. It has the look of a medieval city with ramshackle housing and open sewers, devoid of piped water supplies, public utilities and services like electricity, rubbish collection, etc. In stark contrast, Butembo, its sister city in eastern Congo, stands tall with three storey buildings in blue marble, often traced back to personal fortunes made in gold excavation (Figure 1).

This offering probes the history of urbanisation in these two mineral cities, providing an anthropological interpretation of why their levels of material development and welfare differ. In order to understand why and how the Nande have managed to weather the surrounding violence and build a self-sustaining, prosperous city in the midst of civil war while the Luba have a city deeply scarred by the depredations of civil war, one has to consider the cultural evolution of the Nande's entrepreneurism and the Luba's ethnic extroversion. Indeed, to make sense of the contemporary situation in Butembo and Mbuji-Mayi, one has to integrate the ethnic dimension into the historical unfolding of the continuities and discontinuities of political economic order in the Congo. In the following, I will start with some theoretical remarks before providing historical background on the Congo for an understanding of the nature of the cultural differences between the two cities.

Theoretical observations

The cities of Butembo and Mbuji-Mayi prove that life in the DRC has not died with the so-called collapse of the state. They are urban expressions of new forms of life and organisation emerging from the debris of the state's disintegration; cities of trading networks, emerging elites, new forms of ethnicities, and new local–global connections. Both cities embody the contemporary production of new socialities, differentially exemplified by Nande transnational traders and Luba business enterprises.

In the gap left by the state's retreat, new hybrid forms of governance have emerged among the Nande in Butembo and the Luba in Mbuji-Mayi predicated upon control of gold and diamonds respectively. From the brutal reconfiguration of the DRC state, forms of social organisation have emerged topographically around

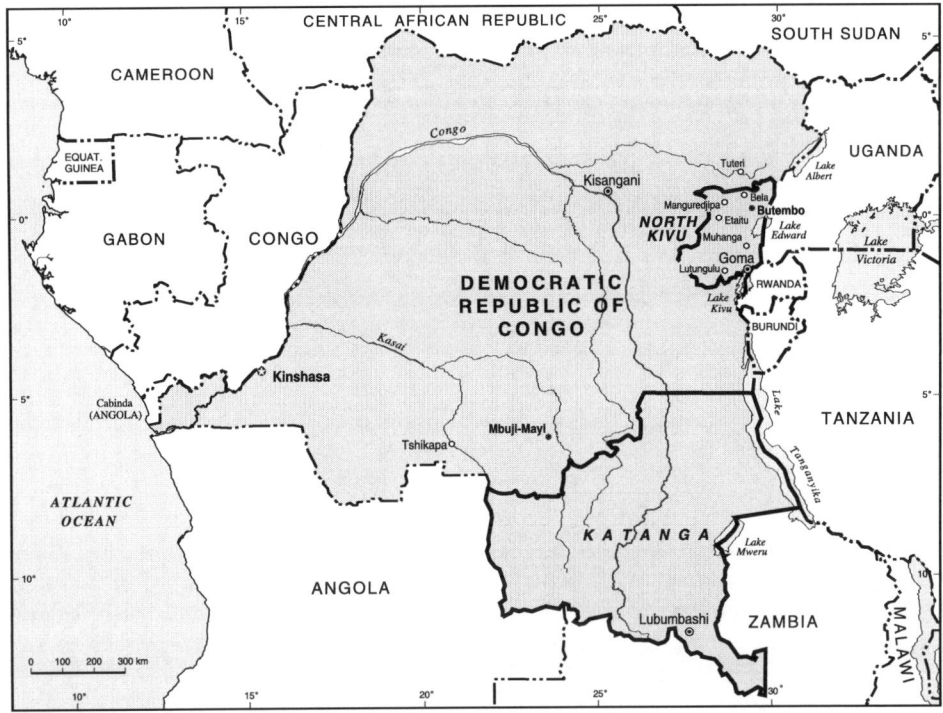

Figure 1. Map of the Democratic Republic of Congo.
Source: Cartography by M. Shand, University of Glasgow, based on United Nations Population Division map, http://www.un.org/Depts/Cartographic/map/profile/drcongo.pdf.

new notably non-national loyalties. In the case of Butembo, the dynamics of global capital accumulation and Nande transnationality collude to produce a reinvigorated sense of ethnic identity in the context of sub-national autonomy within territorial borders that are largely beyond the reach of the authority of the DRC nation-state. Contrasting cities like Butembo and Mbuji-Mayi allow us to explore the historically specific material preconditions of an ethnicity-based reference framework deployed by Nande and Luba traders to extend their economic influence. Probing this phenomenon, we reposition discussion of African state formation into what may appropriately be characterised as the postcolonial 'afterlife' of ethnic differences.

Amidst an abundant literature in African studies sceptical of the influence of ethnic difference, there may be a renewed usefulness and necessity to theorise the salience and continuing production of ethnic differences in a manner that problematises and challenges the notion that ethnicity was merely a divisive invention of colonialism. Mamdani (1996) maintains that ethnicity was 're-invented' under colonial regimes. Capitalism has not transformed African daily life as quickly or completely as modernisation advocates of all stripes thought it might. Mobutu's long reign in the Congo twisted ethnicity to his own needs, imparting a long ethnic afterlife. I argue here that it is important to theorise how ethnicity is being reinvented yet again in cities like Butembo and Mbuji-Mayi.

Indeed, the Nande and Luba's specific histories emphasise the ways in which ethnic particularity congeals over time and becomes deeply embedded in historical and socio-political circumstances. This is not about some kind of essential and eternal 'cultural fixation' of the Nande or the Luba, but rather about the intricacies of their socio-political history, which literally crystallises them as 'Nande' and 'Luba' and contributes to the reification of their respective collective identities as new ethnicities. Nande traders, more so than Luba, form what Tilly (2005) calls 'trust networks' because they are held together by relations of trust and reciprocity. The difference between the two groups is that in the Nande case there is 'strong embeddedness' while the Luba have 'weak embeddedness' (Granovetter 1973).

Trust networks

Theorisation of trust networks can be traced back to two different streams of thoughts. One is rational choice theory, mainly borrowing from neoclassical economics, whereas the other derives from the sociological concept of 'embeddedness' (Granovetter 1973, 1985). In rational choice theory, individual actors' behaviour and utility are not only dependent on individual tastes and interests but also on personal preferences that are heavily but not inevitably determined by their stock of personal and social capital (Becker and Becker 1997). Granovetter (1973, 1985) rejects the narrow rational choice hypothesis as an under-socialised view of society. So too Polanyi (1957) and most recently Tilly (2005), writing about transnational capitalist globalisation, perceive contemporary economic actions and outcomes as closely embedded within social relations. The concept of weak embeddedness is relevant to the level of trust and intimacy involved in a globalised context as compared to a local setting characterised by strong embeddedness. Most importantly, however, one should think about the factors producing change in trust networks. From an internal perspective, one needs to consider what will affect the boundaries, connections, sustenance and internal controls of trust networks. From an external perspective, one must consider what will cause some trust networks to disappear, some to integrate with public politics, others to inform new outgrowths of public politics, and still others to circumvent integration in public politics. My argument is that the move to more global contexts entails significant changes within and between interpersonal trust networks because of the weak embeddedness induced by a global context.

In her ethnographic account of the organisational legacies and informal restructuring strategies of large trade and production networks in Nigeria, Kate Meagher (2010, 23) moved beyond the essentialist tendencies of contemporary networking thinking that reduces African networks to a 'broad cultural logic characterised by popular entrepreneurship, or parochialism and predation'. Instead Meagher uses a more institutional way of reconnecting networks with socio-historical processes, demonstrating how trade, manufacturing and indeed wars give rise to distinctive network dynamics, embedded in particular ethnic, class or gender relations. She stresses that 'the key issue is not the boundary between the official and unofficial spheres, but the distinctive organisational dynamics and power relations that characterise non-formal forms of order' (Meagher 2010, 16). In this way, social networks are historicised and, most importantly, informality is reconceptualised to be not the opposite of formal, but an alternative terrain of regulation operating outside the framework of the state.

To understand Luba and Nande networks and their differential evolution through history, one has to situate their production in the local as well as the global contexts of economic restructuring. Nande trade networks originally were constructed on the basis of local agricultural products such as coffee, potatoes and beans as well as gold, a highly valued international commodity. The Nande have formed networks of trust along the commodity chain of gold from its extraction through its processing to international export. The gold commodity chain includes all the actors that contribute value-added to the final product. Luba trade networks, on the other hand, are centred on diamonds.

Commodity chains

The commodity chain approach emphasises the importance of horizontal coordination among the operators at a given level of the chain as well as the vertical coordination mechanisms between successive levels (cash market, future market, contracts, forward contracts, etc.). I use commodity chains to mean a network of labour and production processes whose end result is a finished commodity (Hopkins and Wallerstein 1986, 17). In other words, commodity chains are traced along a route beginning with raw commodities involving local young men digging minerals, onwards through the military middleman via intermediary buyers and transport operators to the point of exportation where the commodity is shipped to a foreign company that will transform the product into its final consumable form. More generally, commodity chain analysis is concerned with what Arjun Appadurai (1986) called the 'social life of things' or what Kopytoff (1986) refers to as the 'cultural biography of things'.

While Marx focused on the exploitative production relations of finished commodities, Appadurai (1986) is concerned with the sociality of the commodity. In other words, he is interested in social relations induced by the presence of commodities and their circulation. Appadurai (1986, 6) defines commodities as things with a particular type of social potential, distinguishable from 'products', 'objects', 'goods', 'artifacts' and other sorts of things, but only in certain respects and from certain points of view. For Appadurai, a commodity is a thoroughly socialised thing. It is a product intended principally for exchange, and such a product emerges, by definition, within the institutional, psychological and economic conditions of capitalism.

What links value and exchange in the social life of commodities is politics. What is political about this process is not just the fact that it signifies and constitutes relations of privilege and social control, but the constant tension between the existing framework (price, bargaining, etc.) and the tendency of commodities to breach these frameworks (Appadurai 1986, 56). Indeed, such politics can take many forms; politics of diversion and display, politics of authenticity and authentication, politics of expertise and regulation and politics of clientship. Appadurai (1986) has suggested focusing on political and cultural construction of commodity chains as a continuous process of creating and maintaining the framework of rules and meanings within which exchange takes place. His questions then are: how are the paths of exchange laid down? By whom? Whose interest do they serve? What is at stake in battles over commodity paths? As Ferguson (2006) puts it, the questions to be asked are not simply about multi-centric economies or spheres of exchange of wealth, but about commodity path-making in any cultural economic system.

Having outlined the theoretical terrain, we can move to the national context in which the two cities emerged.

Congo in Western literature

The prominent place and depiction of the Congo in western literature on Africa is insidious. From Joseph Conrad's *Heart of Darkness* at the turn of the twentieth century to Barbara Kingsolver's *Poisonwood Bible* (1998) and more recently, John Le Carré's *The Mission Song* (2006), the Congo has swayed cultural representations of critical junctures in Africa's history vis-à-vis the global economy. Conrad's multi-layered novel, considered by some to be racist but by others to indict colonialism and imperialism (Said 1993), was written as the colonial era and the belle époque of global capitalism were reaching their heights (Silver and Arrighi 2003). The Congo assumed a salient position in these upheavals. Polanyi (1957) observed that people in the periphery of capitalism could not protect themselves against the ravaging international imperialism destroying pre-capitalist communities of kinship, neighbourhood, profession and creed – all forms of indigenous, organic societies. It is clear that in the Congo, the demolition of these communities was motivated from the outset by the search for minerals.

In the preceding millennium, indigenous Congo pygmies first welcomed Bantu-speaking groups searching for agrarian livelihoods and metals for their iron and copper tools (UNESCO 1990). Arab traders from the east and later Portuguese merchants from the west entered the Congo in search of ivory and slaves centuries later. In the 1870s, Belgium's King Leopold II laid claim to the Congo as an extension of his personal estate. With the assistance of the Anglo-American adventurer Henry Stanley, he negotiated trade rights for a range of goods, including diamonds, gold and rubber. It is generally accepted that Joseph Conrad's Mr. Kurtz was modelled on Stanley (Wrong 2000). King Leopold's Congo Free State (1885–1908) was a short but a barbaric looting enterprise.

The Belgian Congo (1908–1960) was not so different. Profits from mining enterprises were transferred to the metropolis while the country's education and health facilities remained minimal. Mining companies relied on poorly paid African wage labour whose local purchasing power barely made a dent in the existing hunting and gathering and agrarian subsistence economies of the Congo's vast interior. The mining corporations and the colonial administration paid little attention to 'native' welfare, although considerable capital was invested in infrastructural development in key parts of Congo's vast terrain. This was supplemented by missionaries, notably the Catholic Church, which attempted to provide health and educational facilities in areas dotted around the country in conjunction with their quest for religious converts.

What is critical to note here is that the bulk of mining profits were channelled to the metropolis. One striking example of mineral looting during the colonial period relates to uranium from Katanga used by the Americans to bomb Hiroshima and Nagasaki at the end of World War II. The Americans paid Belgium through the cancellation of its war debts rather than payment to Congo. Overall, Congolese mineral commodity circuits were almost entirely in the hands of foreigners, and in effect pre-empted the Congolese people from realising virtually any of the value from their mineral wealth.

Postcolonial Congo

The Congo gained national independence in the 1960s. Belgium, being a late starter at the colonial game and indifferent to African nationalist aspirations, did not foresee the eventuality of decolonisation in good time. It had trained a pitiful number of candidates to take over as administrative officers. There were only three people with university graduate degrees in the whole country at the time of independence. Nonetheless, the pace of mineral extraction and agricultural commercialisation had already created an urbanised and partially proletarianised population (Peemans 1974). And when demonstrations and riots erupted in 1959, Belgium decided to pull out as quickly as possible, hoping to retain economic and administrative control while pliant politicians enjoyed big cars and other perquisites of power (Moore 2005). This abrupt, unprepared departure was followed by the short-lived rise to prominence of charismatic but poorly educated politicians, notably the ill-fated Patrice Lumumba.

The Cold War pro-imperialist Mobutu dictatorship that followed amounted to an economic hold-up of the country's resources. Mobutu grew rich on the granting of mining concessions to large multinational corporations, and in turn he distributed some of that wealth as patronage through clientelist networks. This sustained the same colonial logic of predation, while bringing the mineral commodity circuit more within the grasp of a small elite within the Congo. Unfortunately they had very little interest in investing the mineral wealth in nation-building and economic development. Mobutu's conspicuous consumption and wealth-dissipating patronage eroded the economic and social fabric of the country over the next 35 years. After he was overthrown in 1997, the neo-Mobutist dictatorship of Laurent Kabila, with a more productionist perspective, used the Congo's mineral wealth as the foundation for a buccaneer industry, enriching himself and his family network at the expense of the Congolese population at large.

The country descended into war in 1998. Within two years, Kabila's Rwandan allies turned on him, and he was assassinated by one of his own bodyguards in January 2001. Since then, the country has been ruled by Laurent Kabila's son, Joseph. Trained at the National Defence University in Beijing, Joseph Kabila has attempted to remove foreign troops while establishing himself as a 'democratically elected' leader of a country roughly the size of Western Europe with over 60 million inhabitants. By the mid-1990s, his lack of control resulted in two massive wars that have economically devastated the Congo, leaving 4 million dead, more than any military conflict since World War II.

During the height of the conflict, several neighbouring countries sent troops to the Congo on the pretext of helping to achieve peace. In the process, the illegal extraction of natural resources in the country increased. Many of the foreign troops have now left, including those of the Sudan, Uganda, Angola and Tanzania, but the Rwandan Hutu forces called the 'Interahamwe' and Congolese militias, known collectively as the 'Mai-Mai' remained. They now occupy the most violently disputed areas of the eastern DRC, concentrating their activities primarily in locations where mining of gold, diamonds and coltan take place. The Rwandan and Ugandan militias are heavily involved in smuggling gold, tantalum and other valuable minerals during their raids across the Congolese border (Perkins 2008). This is despite the fact that the governments of the DRC and its neighbouring countries signed the United Nations-sponsored Lusaka Accord, which ended the first Congo war and formed a United Nations mission in Congo (MONUC) to stabilise the region in 1999. In 2001, the UN issued a report entitled 'Illegal Exploitation of Natural Resources and Other Forms of Wealth of the Democratic Republic of the Congo' (UN Panel of Experts 2001). Despite UN intervention, civil war and exploitation of resources continue with horrific effects on the civilian population at the time of writing.

In the shadowy background of the military conflict and the millions of fatalities in the Congo, the profits of the regional powers and multinational corporations persist. For example, the American Mineral Fields (AMF),[2] an Anglo-American mineral company, became actively involved in the war-torn Congo. It signed a 60/40 deal with Gecamines to extract cobalt tailings from Musonoi River in 2000. Angry with Kabila for ignoring their support to him during the first rebellion in 1996 and signing a deal with their competitor De Beers, AMF became more pragmatic and resorted to backing both sides in the war, lending support to the rebel movement Rally for Congolese Democracy (RDC), in the hope that peace would come along the lines of Lusaka agreement of August 1998.[3]

However, amidst this chaos, Butembo and Mbuji-Mayi function as urban centres and nodes in facilitating the production and export of Congo's mineral wealth. The following sections explore how mineral commodity circuits influence the growth, infrastructural development and welfare of these two cities.

Butembo and the Nande

First-time visitors to Butembo are surprised by its size and prosperity. The main streets are lined with bustling shops, offices and petrol stations. Some buildings have marble surfaces and convey the impression of a wealthy city. Taxis, buses, trucks, motorcycles and *tsicoudo* two-wheeled carts transport people and goods around the conurbation. Butembo is located in a high-altitude tropical forest between the Albertine rift valley in the east and a vast mineral-rich forest to the west, extending to Kisangani in Central Congo. The forest is rich in gold, diamonds and tin.

Origin and growth

Butembo was founded in 1928 and became the centre for the administration of the *Miniere des Grands Lacs* (MGL), which owned the mineral concession in the North Kivu area and dominated early development of the region. Mines were eventually opened up in a number of locales: Manguredjipa (the most lucrative mining area), Etaitu, Tuteri, Lutungulu, Muhanga and Bela (Musekwa 1975). By 1945, some areas employed one-third of the male population in the mines or in some other wage-earning occupation (Packard 1981).

Coffee used to be the main export product.[4] Prior to the civil war, it represented 75% of all national export revenues from agricultural products[5] and was the third major product of export after diamonds and copper. Coffee was exported almost exclusively to Europe (Belgium, France and Switzerland). With high demand for coffee in Europe, its export was at the heart of major capital accumulation in the

Nande territories of Beni-Lubero, especially in the 1970s and 1980s when the world price skyrocketed. During this period, a dozen factories were created in the region for the treatment of coffee.

When the price of coffee declined on the world market, many of these factories went bankrupt. Many coffee traders returned to exporting gold. I use the word 'return' because, in the 1950s and 1960s, gold smuggled into neighbouring Uganda was the first way to obtain the necessary capital to start more legal business activities such as coffee export and vegetable transport within cities. The liberalisation of gold mining in 1982 in Kivu's hinterlands created an opportunity for Nande traders to buy and own gold mines. Since then, they have smuggled gold to neighbouring countries without going through official controls.

Today the population of Butembo is around 720,000, with the Nande constituting approximately 90% of the population. From 1957 to 1998, annual population growth was around 6.5% (Table 1). In the eight-year period from 1998 to 2006, the total population more than quadrupled due to growing insecurity in and around the city, and the expansion of the city limits in 2003 from 25 to 196 square kilometres.

Gold trading foundations

The gold diggers are generally young migrants who sell their gold to intermediary traders. Sometimes they exchange gold for clothes or foodstuffs rather than money. Apart from some individuals who smuggle small amounts of gold to Uganda, most gold from the region ends up in the hands of Nande traders in Butembo.

The intermediary buyers sell the gold to brokers in Butembo who entrust significant quantities of gold to smugglers travelling across international borders by motor vehicle to Kampala, Bujumbura, Mwanza or Nairobi. To cross these borders, they cultivate good relations with customs service agents and immigration control agents. Once they reach their East African destination, the smugglers deliver the gold to a designated exporter who dispatches the gold to Asia or Europe. Payment to the Congolese broker is transferred via a bank or a financial intermediary in East Africa, or alternatively to a person in Asia or Dubai who can deliver the equivalent value of manufactured goods to Butembo. Such merchandise is then delivered by a freight service and then transported by road by the broker's agent to Butembo.

Along the commodity chain, diggers receive the lowest value added while the highest is generally realised at the final destination in Europe or Asia. But Nande traders, who personally sell gold in Uganda, Dubai or Hong Kong, capture more value. During the last 10 years, gold has been exported, principally to Uganda. The difference in prices between Butembo and Kampala justifies this traffic. In 2002, a *tola* (10 grams) cost $100 in Butembo and was sold in Kampala for roughly $115–120.

Table 1. Butembo population by year.

Year	1957	1958	1965	1975	1985	1995	1998	2006
Population	10,916	11,189	19,975	50,921	83,001	137,621	143,493	600,000

Source: Mayor's Office, Butembo.

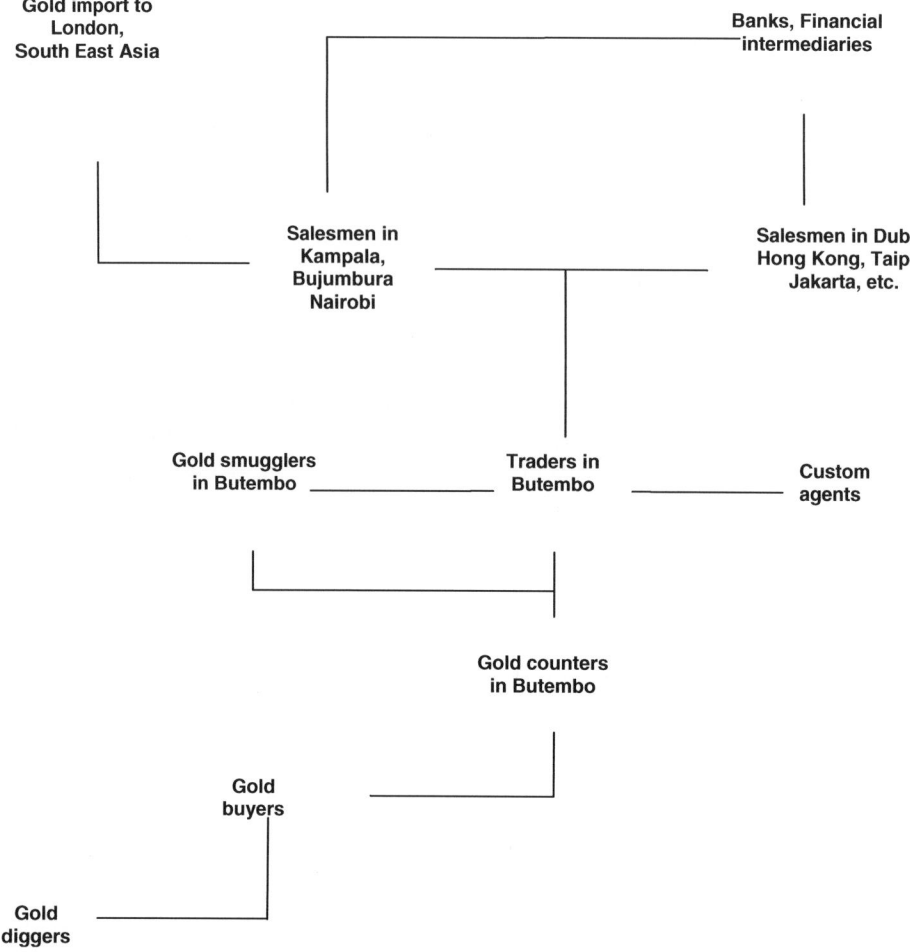

Figure 2. Commodity chain of gold in the Nande region.
Source: Kabamba (2010, 197).

Traders' understated wealth and public investment

The gold trade has dynamised Butembo's local economy. The city's central market area is packed with mini-shops, shops and boutiques selling various items ranging from Chinese plastics to motorcycle spares or textiles from Bangkok or the USA. Around the market, *nganda* (popular bars) have proliferated where Primus beer, roasted chicken or goat meat sustains conversations. During the last five years, many hotels and guest houses have been constructed. Hotel Butembo has provided a modern alternative to the town's oldest hotel, L'Auberge, owned by Mobutu's former governor of the Zairian Central Bank. Most of the new hotels like Ivatsiro and Joli Reve belong to Nande gold traders.

The neighbourhoods in Butembo look like interconnected villages. Vwakyanakazi (1982) contends that the Nande's use of space is dictated more by economic incentives than family needs. The rural pattern of traditional huts built around the home of the headman is a village model that has been discarded. Butembo's compounds are dominated by trade, incorporating small shops where cigarettes, dried manioc, coffee or tea are sold. The same spaces serve as the dwellings for the extended family. Most of the time, traders have two or three buildings, which serve as kitchen, family living space and commercial buildings. However the physical layout is more spacious in the MGL, a neighbourhood named after Kivu's former Belgian mineral company where many traders live. Nonetheless, normally, the size of the buildings rather than location denotes the family's status. Madrabi, a Nande trader who owns Hotel Butembo, has a mansion with 28 bedrooms, eight living rooms and three kitchens. In general rich and poor live side by side in Butembo.

One thing that struck me during my fieldwork was the ordinary way of life of the wealthy in their palatial homes. I was invited by a prominent trader to have dinner and watch the World Cup soccer match final. When I arrived in the dark after sunset, his house was surrounded by beautiful marble statues and an ornate fountain lit by colourful lights. I anticipated a grand dinner of French *foie gras* and *coq au vin*. To my surprise, we ate the normal staple evening meal of *fufu* (cassava), beef and vegetables. Despite their wealth, rich Nande traders remain culturally rooted in their culinary traditions. The simplicity of their daily food consumption is part of a moral economy, which shuns being overly demonstrative about wealth. They live abstemiously in their grand homes, but do indulge in staging spectacular marriage ceremonies and parties.

If one stands on any hilltop of Butembo, the new villas constructed by successful gold traders are visible all over the city. Real estate prices have skyrocketed during recent years. A series of shops called *Gallery Tsongo Kasereka* reportedly cost around US$3 million to build. Most construction materials are imported from China. On the highest hill, an imposing, newly built three-storey rectangular building has a commanding view of the city. This is the mayor's office, which on the south side provides a rural vista of the area where Nande trading activities originated in the 1950s and 1960s centred on vegetable production for export to Kisangani in the Congo and Kampala in Uganda. On the north side, the elegant villas owned and occupied by Nande traders are perched on the site of the former large-scale gold mining company MGL.

Construction of the mayor's office took three years at a time when the surrounding countryside was being devastated by civil war. The building, financed by the Nande traders' association, reflects a sense of civic pride. Furthermore, their mining activities have motivated them to make infrastructural investments in the surrounding countryside. Nande traders are directly financing road maintenance, many taking responsibility for regularly mending up to 50 kilometres. Furthermore, a road tollgate has been constructed to bolster the road maintenance funds. When bad bits of road appear, the traders immediately question their colleague responsible for maintenance of that segment of road.[6] Nande region is now renowned for its good road network, second only to Katanga region, the centre of large-scale mineral exploitation.

In the 1990s, the Butembo traders began taking the initiative in solving the region's transport infrastructural constraints. They started renting Russian-made Antonov planes and air crews from the Ukrainian company *Urga* to export gold and other commodities, paid monthly through international transfers. In 2005, Butembo had five airlines in operation: Air Graben, Cetraca Air Service (CAS), Air Boyoma, Uhuru Airlines and Butembo Airlines. During the last two decades, mobile phones and the internet have also greatly alleviated the transport problems posed by Butembo's remote location. Educational opportunities are also developing. The Catholic University of Butembo, with its six faculties of law, medicine, agriculture and veterinarian, economics, political science and international relations and pharmacy, has continued to function since its establishment 10 years ago by the Roman Catholic bishop of Butembo, Monsignor Kataliko.

The emerging property market and investment in public infrastructure in Butembo illustrates the embeddedness of economic activity in the local community. While capital flight in the rest of the DRC is rampant, in Butembo, houses are built and profits are invested *in situ*. The specificity of Butembo is that there is no legal system of certificate and capital representation to protect the Nande property as would be the case in a western capitalist environment. Security derives from an invisible, informal system of insurance and relationships of trust (Kabamba 2010; Raeymekers 2007; Mirembe Kambale 2004). Nande traders historically made use of their ethnic ties to trade in coffee, beans and potatoes. These practices now facilitate their illegal international gold trading in Asia and the Persian Gulf and other productive ventures. The Nande traders' networks furnish a particularly striking example of successful economic organisation that extends to public investment, providing physical infrastructure, urban cohesion and social welfare.

So, despite the chaotic situation in the DRC, in the absence of an effective national government and state sovereignty and in the face of numerous armed contenders for power, Butembo traders have managed to build a self-sustaining, prosperous city in eastern Congo. As we will see, Mbuji-Mayi in central DRC illustrates quite different social norms alongside their diamond commodity circuit.

Mbuji-Mayi and the Luba

The Luba were comparatively prosperous in the 1990s before the civil wars. During the colonial period, they had developed a legendary opposition to the state, which led to demands for an independent ethnically based territory within the Congo state. Similar to the Nande, the Luba have created an ethnic monopoly over the informal trade of diamonds in their region. The Luba's main city, Mbuji-Mayi, was created on the foundations of the illegal diamond trade (Biaya 1998; Kambayi and Mudinga 1991; Kalulambi Pongo 1997; Missar and Vallee 1997). However, this is only partly true in the sense that there is an array of traditional, regional or national authorities involved in the diamond trade in Mbuji-Mayi to the point that the distinction between illegal and legal transactions is totally blurred.

Mbuji-Mayi is surrounded by the Muya River in the north and Kanshi River in the south. The population of the city is estimated between 1 and 1.5 million and is reported to be the largest city in Africa without any semblance of a piped water system (GeoName 2011). Rapid population growth has exacerbated environmental problems, causing the severe degradation of vegetation cover in the metropolitan area. Wood used to be the only source of energy until Hydroforce, a company which runs the Societé Minière de Bakwanga (MIBA)'s hydroelectric dam recently started providing electricity to residents of the town. The extension of the city has denuded all the arable land in its periphery. Diamond mining, unplanned urbanisation of the living space and inappropriate development of roads along the slopes have catalysed erosion.

Mbuji-Mayi is today regarded as a dilapidated, neglected town. Its appearance does not measure up to the actual resources it generates officially and clandestinely. Despite being located in the region with the highest concentration of diamonds in the Congo and the site of the large-scale MIBA mine, Mbuji-Mayi is one of the least developed cities in the Congo with such important mineral extraction revenues. There is no running water or centrally supplied electricity. Revenues from diamond exploitation are unequally distributed. With the exception of the MIBA neighbour-hood, which has some beautiful houses, the rest of the city has small, ramshackle houses, giving the city the appearance of a vast slum. Local Luba, especially the rich involved in trafficking of diamonds and others who succeed in business, are not inclined to pursue infrastructural investment activities in Mbuji-Mayi and their region more generally. They evince a strong intellectual ethnic diaspora. Money from diamonds has been invested in the education of a young generation who prefer to reside in Kinshasa or abroad in the West.

Mbuji-Mayi is largely populated by the Luba, but with the presence of some important minorities, notably the Songye and Tetela. Women outnumber men not only because they generally live longer, but because accidents befall many men in the diamond mines. During the wars of 1996 and 1998, Mbuji-Mayi remained loyal to the central government of Kinshasa and was defended by Zimbabwean troops who came to the assistance of President Laurent Kabila. The Zimbabwe army sought payment for their services by plundering the city's diamond stocks. It is believed that the preference for the rich to live elsewhere contributed to the defenceless position of the town and its fate at the hands of foreigners. In Butembo, by contrast, the strong presence of Nande 'big men' and their significant investment in the city has made the question of its defence a central concern.

Economically, Mbuji-Mayi is dominated by artisanal and large-scale diamond mining. The majority of the population make their livelihoods through mining or selling services to the miners. Despite the presence of an exceptionally high-value commodity like diamonds, most city dwellers are poor. Diamonds are produced in the environs of the city but the city has exceptionally weak links with the global diamond commodity circuit. Instead the foreign diamond firm MIBA and the central Congolese state monopolise control, channelling the diamond value chain towards Kinshasa, Antwerp and other international diamond centres. Indeed individual buyers in Mbuji-Mayi do make it to Europe or sell diamonds in Kinshasa, but they are the exception rather than the rule. Besides, mine labour constitutes very little of the total value of the diamonds realised in Mbuji-Mayi.

With the downturn in diamond prices, unemployment in Mbuji-Mayi has risen. Many local men migrate to South Africa where they sometimes find work as security guards in the diamond industry. In my interviews with them in 2011 in Johannesburg, these disillusioned migrants pointed to the irony of leaving their rich diamond region to become security guards at a South African company buying diamonds from their region. They bemoan the fact that legal and illegal diamond revenue from Mbuji-Mayi is not being used to develop their city and region, and instead feeds the DRC's state coffers or individual Luba diamond businessmen's conspicuous consumption in Kinshasa and beyond. This ethnic pattern is related to successful Luba traders' aspirations for western education and lifestyles that are unobtainable in Mbuji-Mayi. The few *diamantaires* who invest in Mbuji-Mayi make investments in houses, hotels and import/export businesses. These traders are generally well viewed by the population because they come back and invest at home. However, the diamond trade carries on mainly because it is still considered the 'best game in town'.

Conclusion

The diverging histories of Butembo and Mbuji-Mayi reveal striking contrasts in the way mining profits interact with urbanisation processes. The Nande in the extreme eastern part of the DRC, neglected by the colonial as well as postcolonial state, fended for themselves, and have increasingly used part of their gold profits to support infrastructural development of transport and urban services, in addition to construction of elite, status-enhancing housing. Bound by deep connections with their communities, Nande businessmen act out Granovetter's (1973) 'strong embeddedness', reinvesting in their hometown, bringing their wealth into the urban community for public and private consumption. Giving precedence to strong local links, they nonetheless have managed to profit from a global commodity circuit for their gold.

Historically, the Nande have been primarily agriculturalists. Social cohesion in Butembo can be likened to that of an agrarian society where there is a moral economy in which all benefit from the harvest for the sake of survival and well-being and there is a necessity to be on good terms with other members of the community to ensure the security of one's fields. Among the Nande, there is now a sense that the community is the real owner of their mineral wealth. Wealthier members of the society feel obliged to make an effort to redistribute some of their wealth to others in the city in the form of infrastructure provisioning. These practices are absent in the Luba region.

First, the Luba's social fragmentation in the aftermath of the slave trade created high levels of insecurity that pushed individuals to seek refuge and protection from external patronage. Individuals migrated in search of safety. Eventually, during the Belgian colonial period, individual Luba sought security through education and employment in the Catholic missions, the administration and the mineral companies. Here there is no sense of community comparable to what could be found amongst the Nande in Kivu. Local Luba diamond businessmen are mostly established in the capital city, Kinshasa. Within the diamond commodity circuit, they largely ignore the local level and pursue far stronger linkages with national and international nodes. Their relation to the community in Mbuji-Mayi is an extractive rather than a constructive one. This fact has become etched in the city's poor physical infrastructure, arrested growth and relative poverty.

In brief, a Luba diamond trader has the tendency to 'grab and go', while the Nande trader seeks to 'dig then develop'. The Nande are far closer to Polanyi's (1957) sense of economic transformation. It is interesting that this strategy has been pursued in relative isolation from the DRC nation-state. The Nande's strong local ties were a survival strategy that succeeded against the odds and has gone on to provide its own impetus for urban innovation.

One of the explanations that Luba traders give concerning their extractive relations vis-à-vis their hometown is that the social pressure exerted on them to redistribute is too strong and intimidating because of the depth of poverty in the town. The majority of city dwellers are unemployed and extremely poor. Diamond businessmen feel more comfortable and safe in the anonymity of Kinshasa and

European capitals. They avoid encounters with Luba in their hometown, following a non-embedded, individual rather than collective strategy for attaining economic success.

As argued above, there are many cultural and historical elements that inform the uneven processes of urbanisation in the DRC's mineral-rich cities. Meagher's (2010) emphasis on exploring the institutional history of indigenous economic networks and the ways these networks have been restructured by changing political and economic conditions in relation to the changing nature of the nation-state is vital to an understanding of the difference between the two contrasting paths the Nande and the Luba have taken. So, too, Appadurai's (1986) concept of a social commodity provides a meaningful way of tracing the commodity circuit and the cultural trajectory of people who live in the mineral-producing area. The social and economic influence of Butembo as a mineral-rich city has been enhanced by the collective meaning culturally imputed to the gold. The value of the gold is realised by the individual trader and, to some extent, its value is embedded in the infrastructural development of the city. By contrast, in Mbuji-Mayi, diamonds are far more ephemeral and individualised. Luba traders procure diamonds for onward sale in an international commodity circuit in which Mbuji-Mayi is quickly made inconsequential with virtually no positive spill-over effects for the local urban population. In Butembo, the trading agents are grounded and motivated to invest their diamond wealth for individual and collective welfare, whereas in Mbuji-Mayi diamonds, as a social commodity, are whisked away for the benefit of traders and consumers elsewhere.

Notes

1. Coleman (2005) notes that the number of deaths in the Congo conflict amounts to a death toll higher than the combined estimated deaths from the conflict in Darfur as June 2005 (400,000), the 2004 tsunami (937,000), and all American deaths in every war fought since 1776 (1,540,665). The figure of 4 million is provided by International Rescue Committee (IRC) in its 2006 Annual Report (http://www.theIRC.org). For IRC, only 10% of the deaths were directly caused by bullets. The rest were indirectly related to war and the effects of infectious disease and malnutrition. Many NGOs contested IRC figures as exaggerated and point to important methodological problems in the IRC study. Doctors Without Borders, for example, put the death toll of the Congolese conflict between 1 and 2 million (Autesserre 2010). On the uncertainty surrounding the IRC's statistics see Strategy Page (2006).
2. 11% is owned by Union Miniere and 89% by Anglo-American.
3. This situation is not at all new in the history of warfare. In World War II, the Ford Motor Company was building engines for the German Fokker planes used by the enemy's air force.
4. Eastern Congo was among the principal coffee-producing zones in the country.
5. Republique du Zaire, Conjuncture Economique, Departement de l'Economie Nationale, Industrie et Commerce Extérieur (1985, 41).
6. Most of the business elite are Protestants espousing a community-minded ethic.

Note on contributor

Patience Kabamba is Assistant Professor of International Studies at Marymount Manhattan College and Senior Associate Researcher at the University of Johannesburg. His theoretical interests are the dynamics of conflict, new state formations, transnational trade networks, ethnicity and global political and economic governance. He has extensive ethnographic experience in the Democratic Republic of the Congo as well as in Rwanda, Burundi and Uganda. He has published articles on ethnicity, rebel-controlled territories, economic empowerment and modes of political production in the *Canadian Journal of African Studies, Congo-Afrique, Anthropological Theory and Africana* and a book entitled: *Business of civil war: New forms of life from the debris of the state* (CODESRIA & Indiana University Press, 2012). He can be contacted at: pkabamba@mmm.edu

References

Appadurai, A., ed. 1986. *The social life of things: Commodities in cultural perspective.* Cambridge: Cambridge University Press.

Autesserre, S. 2010. *The trouble with the Congo: Local violence and failure of international peacebuilding.* Cambridge: Cambridge University Press.

Becker, G.S., and G.N. Becker. 1997. *The economics of life.* New York: McGraw-Hill.

Biaya, T.K. 1998. Le pouvoir ethnique: Concept, lieux de pouvoir et pratiques contre l'etat dans la modernité Africaine: Analyse comparée des Mourides (Sénégal) et Luba (Congo-Zaïre). *Anthropologie et Sociétés* 22, no. 1: 105–35.

Coleman, S. 2005. Congo's conflict: Heart of darkness. *Beliefnet Newspaper,* June 2, 2005.

Conrad, J. 1999. *Heart of darkness.* Plymouth: Broadview Press.

Department de l'Economie Nationale, Industrie et Commenrce Exterieur. 1985. *Republique du Zaire: Conjuncture Economique.* Zaire: Department de l'Economie Nationale, Industrie et Commenrce Exterieur.

Ferguson, J. 2006. *Global shadows: Africa in the Neoliberal world order.* Durham, NC: Duke University Press.

GeoName. 2011. Geographical database. http://www.geonames.org/ (accessed January 21, 2011).

Granovetter, M. 1973. The strengh of weak ties. *American Journal of Sociology* 78, no. 6: 1360–80.

Granovetter, M. 1985. Economic action and social structures: The problem of embeddedness. *American Journal of Sociology* 91, no. 13: 481–518.

Hopkins, T., and I. Wallerstein. 1986. Commodity chains in the world economy prior to 1800. *Review* X, no. 1: 157–70.

Kabamba, P. 2010. Trading on war: New forms of life from the debris of the state. PhD diss., Columbia University.

Kalulambi Pongo, M. 1997. *Etre Luba au XXème siècle.* Paris: Karthala.

Kambayi, B., and M. Mudinga. 1991. *Le 'citancisme': Au coeur de l'évolution de la société Luba-Kasai.* Kinshasa: Saint Paul.

Kingsolver, B. 1998. *Poisonwood bible.* New York: Harper Perennial.

Kopytoff, I. 1986. The cultural biography of things: Commoditzation as process. In *The social life of things,* ed. A. Appadurai, 64–91. Cambridge: Cambridge University Press.

Le Carré, J. 2007. *The mission song.* New York: Little, Brown and Company.

Mamdani, M. 1996. *Citizen and subject: Contemporary Africa and the legacy of late colonialism.* Princeton, NJ: Princeton University Press.

Meagher, K. 2010. *Identity economics: Social network and the informal economy in Nigeria.* Oxford: James Currey.

Mirembe Kambale, O. 2004. Autour de l'économie informelle en période de guerre en République Démocratique du Congo, *Parcours et initiatives, Revue interdisciplinaire de l'Université Catholique du Graben,* no. 1: 18–36.

Misser, F., and O. Vallée. 1997. *Les Gemmocraties: L'economie politique du diamant Africain.* Paris: Desclée De Brouwer.

Moore, D. 2005. Banging heads together with velvet gloves: On transitions in the Democratic Republic of the Congo. Paper presented at DRC Symposium: Perspectives on the DRC Transition, Institute for Global Dialogue, Burger's Park Hotel, Pretoria, May 30, 2005.

Musekwa, M.K. 1975. Evolution du commerce dans le centre de Butembo (1928–1958). Honor's Work, Institut Superieur de Bukavu, DRC.

Packard, R.M. 1981. Social change and the history of misfortune among Bashu of Eastern Zaire. In *Explorations in African systems of thought*, ed. I. Karp and C. Bird, 237–66. Bloomington: Indiana University Press.

Peemans, J.P. 1974. Capital Accumulation and State Policy: the Case of Congo. In *Colonialism in Africa*, Vol. IV, ed. Peter Duggan and Lewis Gann. Cambridge: Cambridge University Press

Perkins, J. 2008. *The secret history of the American empire.* New York: Plume Publishers.

Polanyi, K. 1957. *The great transformation.* Boston: Beacon Press.

Raeymaekers, T. 2007. The power of protection. Governance and transborder trade on the Congo-Uganda frontier. PhD diss., Ghent University.

Said, E. 1993. *Orientalism.* New York: Vintage Books.

Silver, B., and G. Arrighi. 2003. Polanyi's 'double movement': The belle époques of British and U.S. hegemony compared. *Politics and Society* 31, no. 2: 325–55.

Strategy Page. 2006. The true cause of death in the Congo. *Strategy Page*, January 12, 2006. http://www.strategypage.com.

Tilly, C. 2005. *Trust and rule.* Cambridge: Cambridge University Press.

UNDP. 2011. Human Development Index. http://hdr.undp.org/en/statistics/ (accessed June 3, 2012).

UNESCO. 1990. *General history of Africa*, vol. 2, abridged by G. Mokhtar. Oxford: James Currey.

UN Panel of Experts on Illegal Exploitation of DRC Mineral Resources. 2001. Report on illegal exploitation of DRC mineral resources and other forms of wealth in the Democratic Republic of the Congo. http://www.un.org/news/dh/latest/drcongo.htm.

Vwakyanakazi, M. 1982. African traders in Butembo, Eastern Zaire (1960–1980). A case study of informal entrepreneurship in a cultural context of Central Africa. PhD diss., University of Wisconsin-Madison.

Wrong, M. 2000. *In the footsteps of Mr. Kutz: Living on the brink of disaster in Congo.* London: Fourth Estate.

Angola's planned and unplanned urban growth: diamond mining towns in the Lunda Provinces

Cristina Udelsmann Rodrigues and Ana Paula Tavares

Mining towns in Angola have followed a different growth trajectory from urban agglomerations elsewhere on the continent. Colonial mining cities were treated as regional strategic locations mainly under the direction of mining companies, with an orientation towards natural and human resources management and planned urbanisation. As Angola became engulfed in civil war after independence in 1975, urban planning and control fell into disarray, which led to rapid and unplanned urban growth as rural populations fled the insecurity of the countryside. In the provinces of Lunda Norte and Lunda Sul, where diamond mining dominates economic activity, new 'unplanned' clandestine towns appeared during the civil war. Since the advent of peace in 2002 and the cessation of war-induced urbanisation, Lundas' towns have resumed their status of government and mining company-dominated settlements, where control and planning prevail. However, improved infrastructural conditions prevail alongside new forms of social and economic exclusion. This article traces urban growth and welfare in Lunda, analysing the impact of its divergent growth pattern on the urban population.

An understanding of contemporary Angolan urbanisation in the Lunda provinces hinges on comprehension of the intimate triangular relationship between urban change, mineral exploitation and conflict. Oil and mineral wealth is the dominant economic influence in the national economy. In the first decade of the twenty-first century, offshore oil and diamond exploitation propelled Angola to achieve the world's fastest growing GDP at 11% in 2010. Part of this astounding rate relates to Angola's exceptionally low starting position in 2002, rising from a no-growth economy encumbered by the accumulated devastation of over 30 years of civil war. Mineral and oil wealth has allowed Angola to make a remarkably quick economic turnaround, accompanied by some surprises in the spatial pattern of urbanisation.

During the prolonged war period (1975–2002), government-controlled offshore oil profits were channelled to the cities, especially the capital, Luanda, where the majority of the national population had fled. Diamonds, concentrated in upcountry locations particularly in the north-eastern Lunda provinces, provided resources to the opposition guerrilla movement UNITA[1] and stimulated population agglomeration in mining areas (Figure 1).

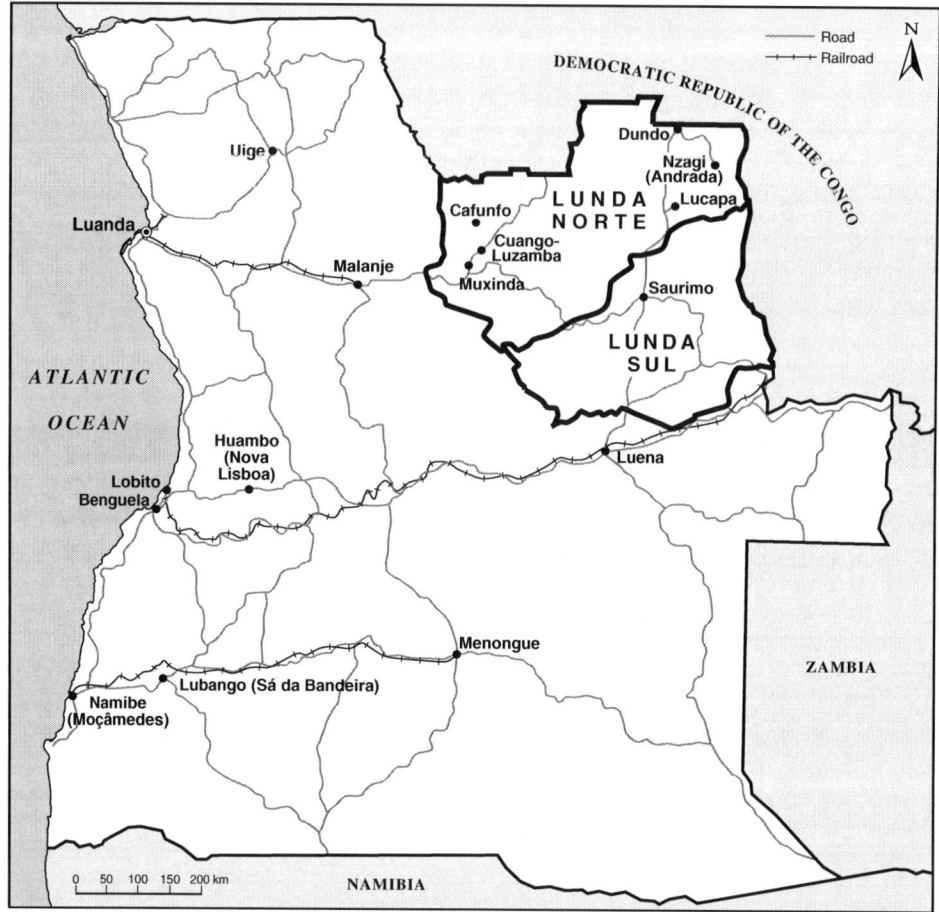

Figure 1. Map of Angola and Lundas Provinces.
Source: Cartography by M. Shand based on United Nations Cartographic Unit 2008 map.
http://www.un.org/Depts/Cartographic/map/profile/angola.pdf.

Control of the diamond areas of the Lundas[2] determined the type of colonial settlement in these inland areas. Mineral resources, particularly diamonds, became associated with the territorial competition of warring factions throughout the civil war, dictating the positions of conflicting sides and settlements. From 2002, the Lunda provinces regained the attention of government due to the pressing post-war need to regulate mineral resources and plan settlement expansion.

Despite the 'unplanned' nature of Angola's urban peripheries (Jenkins, Robson, and Cain 2002), the country's major colonial cities have been structured by the economic, political and social motivations of urban planning strategies. The same applies to the cities in the Lundas. However, there are important differences in these provinces, related to the civil war and subsequent unplanned growth. While the colonial Lunda cities' growth was from the very beginning determined, planned and under surveillance of the mining company Diamang and the colonial authorities, the post-independence government was unable to maintain control of the provinces.

The Lunda provinces became the stage for military interventions and competition for resources. A large proportion of the local population sought refuge in the region's major cities or in the coastal ones. Meanwhile, spontaneous concentrations of both civilian and military populations in search of diamonds at specific mining sites catalysed rapid and unpredictable growth on new sites.

Urban growth in the Lundas is intrinsically related to major socio-political changes in the country since the sixteenth century: colonisation until 1975, civil war between 1975 and peace from 2002. The social history of mining is entwined with urban growth and has a similar chronology: the colonial corporate regime, the lawless presence of wartime fortune seekers, and the current quasi-sovereign state of corporate mining (Calvão 2011) combined with clandestine *garimpo*.[3] This contribution begins by tracing the overall pattern of urbanisation in Angola, followed by a discussion of the role of mining and conflict as an impetus to urban growth and upheaval in the Lundas. The third section dissects distinctions in operations between large-scale corporate mining and small-scale artisanal mining, and their different forms of impact on urbanisation. Finally, our analysis focuses on the conditions that have led to contrasting patterns of wealth and poverty in the 'old' and the 'new' urban settlements and their link to post-war regulation and planning.

Urban growth patterns in Angola

The country's colonial urbanisation history began with the establishment of the port of Luanda to serve the trans-Atlantic slave trade in the sixteenth century. Urban settlement remained a largely coastal phenomenon throughout several centuries of colonial rule. However, upcountry towns emerged and slowly evolved with the commercial development of hinterland export crop production and mineral exploitation during the twentieth century.

Prior to the arrival of the Portuguese, there were densely populated settlements connected with important chieftainships in Lunda (Mussumba), Kongo (Mbanza Kongo) and Huambo (Ombala) (Carvalho 1890–1894). These pre-colonial towns lost their political and administrative importance with colonisation.

Today's urbanisation in Angola has its origins primarily in colonial patterns of settlement. Some grew from trading posts and colonial expansion posts that combined commercial, religious and defensive military characteristics. The European slave trade to the Americas was associated with the establishment of the most important coastal urban centres, namely the capital Luanda, as well as Benguela. In the mid-nineteenth century, fewer than a million inhabitants lived in the colonial districts under Portuguese administration (Dias 1989, 243). Penetration of the interior came with the development of agriculture, industry and transportation networks, including construction of major railway lines, and an increased white settlement policy in the second half of the twentieth century that contributed to the emergence of cities like Sá da Bandeira (Lubango), Nova Lisboa (Huambo) and Malanje that grew in proportion to their administrative functions. The ports of Lobito and Moçâmedes were also substantially developed in the second half of the twentieth century, though traffic volumes through these southern routes never equalled that of Luanda. By the first half of the twentieth century, Angola had an upcountry urban network, which remained highly dependent on road and rail connections to the main coastal ports dominating the urban hierarchy (Table 1).

Table 1. Population in major Angolan cities.*

City	Date founded	Population 1950	Population 1970	Population 2010
Luanda	1576	141,647	475,328	5,230,000
Benguela	1617	14,690	40,996	177,169
Namibe/Moçâmedes	1840	8,576	12,076	79,047
Lobito	1842	23,897	59,528	186,871
Malanje	1852	9,473	31,599	162,877
Lubango/Sá da Bandeira	1885	11,654	31,674	257,894
Huambo/Nova Lisboa	1912	28,296	61,885	315,147

*It should be noted that general population data have not been collected since 1970. A national census is planned for 2013. In the meantime, estimates and partial statistics have been generated by both the government and international agencies, especially the United Nations. The best reliable data, congruent with these estimates are available through E-Polis, showing the massive growth of cities in Angola, especially the coastal ones.
Source: Amaral (1962); E-Polis (2010).

Since it was founded, Luanda has consistently been the most populous and important city in the country, combining administrative, commercial, financial and industrial functions of national relevance. While all coastal urban centres grew during the civil war, Luanda amassed remarkable numbers of people, amounting to nearly one-third of Angola's entire population.[4]

Through the years, settlement in the country was planned according to economic and administrative interests. In the 1970s, urban plans were implemented in already established urban centres. As in other African countries, urban growth was not triggered by industrial growth (Bryceson 2006). Colonial government authorities 'allowed' this unplanned growth, as long as it was kept in the outskirts of the city. The 'unplanned' squatter areas, the *musseques* periphery, grew regardless of attempts to control it, propelled by the agency of African dwellers seeking livelihoods and residential space (Simone and Abouhani 2005; Potts 1997). In major cities like Luanda or Benguela some of the spontaneous *musseques* were, late in the colonial period, subject to planning (the *bairros indígenas*). After independence, with massive migration to the cities, the peripheries' growth accelerated, becoming less constrained and even less anticipated and provisioned in terms of services.

The dynamics of growth of other urban centres in Angola were better handled from the beginning. Some of these urban centres had administrative functions,

Table 2. Population in the Lunda provinces.

Province	Urban area	1960	1970	1980	1990	2000	2010	2020
Lunda Sul	Saurimo	3,100	12,901	20,815	33,584	54,186	83,470	112,754
	Dundo				10,300	23,400	47,120	70,840
	Lucapa			5,200	13,500	34,900	30,000	38,000
Lunda Norte	Cafunfo				7,500	14,600	26,010	37,420
	Nzagi (Andrada)				7,500	14,500	25,960	37,420
	Cuango-Luzamba				5,100	8,200	12,414	16,628
	Muxinda					7,200	10,516	13,832

Source: E-Polis (2010).

whereas others emerged as towns to service the needs of settler populations involved in agricultural export, the *colonatos*, and in the Lundas, populations involved in mining activities. In the north and north-eastern part of the country, the Lunda cities were comparatively less dynamic throughout the colonial period. Lunda provinces, with their highly dispersed populations, were perceived to be too distant from sea and communication routes and virtually deserted. The construction of the railway in the first years of the twentieth century contributed considerably to the expansion of Malanje, the administrative capital city of the Lundas until the 1920s. Malanje gained importance as an urban centre, attracting Portuguese settlers who found a way of making a living trading agricultural products from the coffee and sisal plantations of the surrounding area. This income, along with a rise in sisal prices, was a great boost to Malanje's urban growth during the 1940s and 1950s (Amaral 1962). Cotton replaced sisal when the latter's prices began to decline in international markets.

In the 1970s, diamonds from Malanje and from the neighbouring Lundas caught the attention of investors, the state and the local population. At that time, only Saurimo, which combined mining investment with a strategic location on the main national road axis and good connections to Malanje and the Benguela railway, grew significantly in population and infrastructure. Dundo, in the north, was experiencing more focused diamond-led growth. The company in charge of diamond mining in Lunda, Diamang, fostered urbanisation in the form of enclosed residential compounds for mine workers, not only to control mining activities and staff, but also to better manage the local and displaced workforce as well as the rural population settling around the mine (Cleveland 2008a, 2008b).

Despite greater regulation in the Lundas cities historically,[5] the situation during the post-independence civil war was similar to other cities in Angola. They too absorbed a massive population from the countryside, and their peripheries grew uncontrollably. However, the main difference was that unplanned settlements, which grew up in new areas, were heavily associated with the *garimpo*. These were the only new, non-colonial urban locations that emerged in the region during the country's protracted civil war.

Interplay of mining, urbanisation and conflict

Colonial formalism and large-scale mining influences

The first colonial settlements in Lunda were militarily secured to start farming activities and attract settlers,[6] while not disregarding the possibilities of finding gold or copper in the area.[7] When diamond mining started in the early twentieth century, it influenced the location of towns and paved the way for the province's main urban centres.

The first discoveries in Lunda province took place in 1912 in the Mussalala River by *Société Internationale Forestière et Minière du Congo* (Forminiére), close to the present-day border of Angola with the DRC.[8] PEMA (*Companhia de Pesquisas Mineiras de Angola*)[9] was founded that same year.[10] The establishment and development of large-scale colonial mining sites was actively promoted by the state. Diamang, the state-owned Portuguese company with capital from De Beers and other European companies, was set up in 1917.[11] The first open cast mines were opened in 1919 in Cavuco, Camimanga, Cassanguidi and Luaco (Pereira, Rodrigues, and Reis 2003). Alluvial diamonds dominated the mining activities during the colonial period, whereas kimberlite pipes began to be exploited after independence.[12] Colonial diamond digging in the region and the development of mining infrastructures spurred the provision of housing and medical and educational services for the miners and their families providing the nucleus for new provincial cities and towns.

The pre-independence data shows that the mining industry employed about 18,000 individuals. Besides the Portuguese and Angolan qualified workers with managerial and technical responsibilities, the industry relied on forced labour,[13] recruited locally in the first instance, but because the area was so sparsely populated there was need for the import of significant numbers of labourers into the region.[14] Throughout the colonial period, urbanisation in the Lunda mining regions had largely derived from mining companies' planned initiatives and needs. By independence, urban centres associated with formal mining such as Dundo[15] or Lucapa were already well established.

Large-scale mining, notably the Cambulo,[16] Citatu and Kamissombo mines, gave impetus to the towns of Lucapa, Dundo and Andrada (now Nzagi). Saurimo,[17] now the 'diamond city' and capital of the Lunda Sul province, is associated with the Catoca mine,[18] though its growth entailed other aforementioned factors. Investments in urban infrastructure were also initiated and promoted by successful traders in the diamond business and associated trades. The government and corporate diamond mining policy was simultaneously one of urban development linked to the strategic control of the diamond-rich region.

Garimpo, blood diamonds and reconfigured urbanisation: 1975–2002

Three years after national independence, the colonial Lunda district was divided into Lunda-Norte and Lunda-Sul provinces with capitals in Lucapa and Saurimo, respectively. In the nearly 30 years of civil war that followed, mining cities and sites were subject to intense dispute between the mining companies state-regulated by Endiama[19] and the UNITA guerrillas. Their contestation deeply affected the settlement of population in the region. While living conditions outside the main cities were extremely difficult due to the war, a significant number of military and garimpeiros were attracted to the mining areas.

The Angolan civil war was fought mainly in the rural areas, forcing huge segments of the population to take refuge in the country's main coastal cities, especially Luanda. People fled to the capital from all of the country's provinces (Amaral 1962; Rodrigues 2007). From half a million inhabitants before independence, the capital's population trebled in size over the next two decades. Coastal Lobito, and Benguela as well, showed higher population growth during the war, compared to the hinterland cities of Malanje, Huambo and Lubango. Concurrently with the large numbers trying to evade the ill effects of the war, cities like Malanje or Huambo experienced fluctuating populations due to sporadic, and often intense, disputes over their control by the MPLA-led[20] government as opposed to UNITA. In Saurimo, civil war forced tens of thousands of rural dwellers to seek refuge from war-torn areas of the interior and settle in a city planned for 18,000 inhabitants. Saurimo was cut off from the rest of the country as both armies controlled road traffic. The government held the cities while the insurgents operated in the countryside (Malaquias 2007). Besides destabilising the countryside, UNITA's strategy throughout the civil war focused on securing areas rich in natural resources. This gave them a rationale to drive the rural population away, pushing them into the government-controlled urban centres. In this way, the UNITA elites gained ample space to 'enrich themselves without the political and administrative costs of governing' (Malaquias 2001, 317).

Lundas' cities also integrated new spontaneous and precarious forms of housing and infrastructure in their unplanned *musseques*. With the onset of the war, the mining cities' spatial and urban structure radically altered. Indeed, the population of the *musseques* shifted from the periphery into more central neighbourhoods, as observed in other Angolan cities (Rodrigues 2009). The civil war also triggered a massive exodus from the formal mining areas. Both Portuguese and contracted workers were forced to seek personal safety, the former going back to Portugal and the latter moving to the cities.

The halting of large-scale mining production created conditions for the mushrooming of small-scale mining sites with highly mobile and unstable work-forces. Artisanal mining sites proliferated amidst the deprivations of war. With the collapse of formal mining employment, *garimpeiros* (small-scale miners) and unlicensed diamond dealers arrived from all provinces of Angola as well as other neighbouring countries. They were spurred by UNITA's recruitment of mine labour especially from villages along Angola's border with the DRC and Zambia (Le Billon 2008; Marques and Campos 2005).

The Angolan authorities' attempts to control these movements met with little success as cross-border migration in the Lundas surged during the 1980s. Diamond exploitation was increasingly controlled by UNITA involving a hierarchical arrangement of local and foreign *garimpeiros* associated with the UNITA military. The sale of what came to be known as 'blood diamonds'[21] were manipulated by a complex network of buyers, which included licensed *comptoir* buyers and unlicensed, illicit buyers (De Boeck 2001; Malaquias 2007). A number of diamond buyers settled on the Congo side of the border, building a complex network of exchange. Some *garimpeiros* also managed to establish varied types of informal businesses on the basis of their earnings.

The population living in the mining areas in the Lundas consisted essentially of younger men, many of them immigrants, dedicated to the *garimpo*. There was a variable number of military, mostly men, of different ranks, traders and businessmen supplying both the military and the *garimpeiros* and a few sponsors and diamond buyers operating locally and internationally through extensive networks. While many young men came in search of '*lubóia*', the big stone, the salvation stone (Tavares 2010), they were in fact partaking in the region's growing impoverishment.

Urbanisation was spurred by artisanal mine production sites and all the intermediary activities of the diamond business in the Lunda provinces (De Boeck 2001; Vlassenroot and Büscher 2009). Informal *garimpo* mining contributed to the creation of new urban settlements smaller than the colonial planned cities. These locations gradually acquired an urban character resulting from the population expansion and economic dynamism associated with diamond mining and contrasted with the government-controlled cities that were almost permanently under siege. However, the civil war and miners' constant contentious claims at mining sites precluded the development of urban infrastructural development. As such, the new towns were largely composed of miners and traders living in urban circumstances devoid of infrastructural investment, let alone urban planning.

With this type of population concentration in mining areas, places formerly unpopulated or sparsely populated became crowded and animated as exemplified by Cuango (estimated population 24,000), Cafunfo (100,000), Luremo (16,800), Muxinda and Xa-Muquelengue (USAID 2008, 6). Cuango, associated with the Cafunfo mine and located on the road to Lundas, Malanje and Luanda, emerged as a large city in Lunda Norte. The settlement had been the site for the illicit traffic of diamonds by foreigners since the nineteenth century (Ferrão 2005). Towards the end of the colonial period in the 1960s, commercial diamond exploration began in the Cuango valley, catalysing growth on the basis of non-government investments, without planning or adequate infrastructural development to accommodate the expanding population. This typified the nature of urbanisation in the region.

In effect, when the civil war period is compared with the preceding colonial era, patterns of population attraction and repulsion are striking. The colonial period was characterised by the repulsion of forced labour in the rural areas and the consequent attraction of rural migrants to the cities. Similarly, during the civil war coastal urban centres, and especially Luanda, acted like magnets for populations repulsed by war from their rural home areas. But what differentiates the Lundas is that new urban mining sites emerged amidst the war and massive migration to government-managed cities. Either by force or opportunity, pockets of artisanal diamond miners gave impetus to sizeable population concentrations and commercial activities.

Peace and the remaking of Lundas' cities

With the end of war in Angola, a massive return of the displaced migrant populations to their homelands was expected, but in fact there was no deluge of people opting to move back to the countryside (ADRA 2004; Costa and Rodrigues 2008). Inland cities that managed to recover from wartime destruction began to flourish as trading opportunities proliferated and government investments were initiated. Previously, safety had resided in the coastal cities and in Luanda during the war, but now prosperity became the people's beacon.

There was growing prosperity in diamond mine production areas. Angola is currently the world's fifth largest diamond producer by value, and its gems are coveted for their size and purity (Allen 2010). Diamonds' weighting in Angola's GDP grew from 1.4 to 46.9 billion Kwanzas between 1999 and 2003 (Isaksen, Tvedten, and Ilinga 2006, 3). Angolan mining today consists of large-scale industrialised mining operations, on the one hand, and unregistered diamond digging by small-scale miners in locations beyond the control of the state and mining companies, on the other. Jobs have expanded in the formal mines and associated activities (outsourcing companies, commerce, services) while in the informal *garimpo* sector activities became more discreet, In Cafunfo, for instance, over $1 million was estimated to be circulating daily in *garimpo* circuits (Marques and Campos 2005, 12). Recent growth of other economic investments in the Lundas and opportunities in the public sector provide diversified employment besides mining, but mining retains economic centrality.

As sources of income for the population broaden, the economic dynamism of the urban areas strengthens. The Angolan government is regaining some degree of control over urban management and mining, including efforts to crack down on the immigration of artisanal miners from neighbouring countries. The state has resolved to address issues of urban expansion and development. New city planning regulations are aimed at checking urban growth and improving the spontaneous, unplanned wartime settlements. Recent reconstruction in Saurimo is visible in the poor, unplanned areas surrounding the city that house an estimated 80,000 inhabitants.

Today, the Lundas are once again highly dependent on mining companies' investment in infrastructure. Catoca mine, currently the fourth biggest diamond mine in the world, and Camatchia-Camagico mine, that started its operation in 2007, are attracting migration to Lucapa and Luó. Mining companies are now facing pressure to meet their 'corporate social responsibility', which increasingly is defined in terms of infrastructural provisioning for the population around the mining site.

Migration of informal *garimpo* is another story. Over the past decade, government policies and actions have been directed at reversing a growing trend for immigrants from neighbouring countries to settle in the Lundas provinces. During *Operação Brilhante* in 2004, over 100,000[22] foreign citizens were forcefully expelled and repatriated, 90% of whom were from the DRC (Marques and Campos 2005, 9). Today, the source of artisanal mine labour is mainly local or from other Angolan provinces. However, despite numerous government repatriation exercises, a significant portion of *garimpeiros* are still of foreign origin (mainly from Senegal, Côte d'Ivoire, Sierra Leone and DRC).

Simultaneously, government investment in urban centres and infrastructure since the cessation of war in 2002, together with improving world diamond prices, have attracted Angolans from other provinces to administrative posts in Saurimo, Lucapa and Dundo. Urban growth in Lundas is currently more orderly than before. *Garimpo* that formerly prevailed in the region under force and intimidation is increasingly subject to state regulation, despite the new laws of 2011 on artisanal mining that legalised it under certain conditions. Meanwhile, government policies are endeavouring to attract young people to educational opportunities and administrative jobs.

Who benefits from Lundas mining wealth?

Impacts of formal and informal mining on urbanisation

Angolan diamond mining continues to be concentrated in Lundas, exerting a profound impact on population dynamics and urban growth (Pereira, Rodrigues, and Reis 2003). All the cities and towns of the Lunda provinces are affected by the development and expansion of formal company-based extraction by concession companies[23] and/or informal, small-scale unlicensed *garimpo* miners (Dietrich 2000). Different configurations of formal, informal, state and private agents are generating diverse urban poverty and wealth outcomes. Large-scale mining affords bigger and more permanent formal settlements as opposed to small-scale mining's makeshift settlement, urban infrastructure, housing and service provisioning arrangements for the incoming population. The result is that unplanned towns within informal diamond mining are tending to disappear.

The Lundas cities that have been historically associated with large-scale mining managed, despite the hardships of civil war and a massive influx of population and informal types of settlement, to retain a semblance of basic urban structure, which is now being strengthened by government reconstruction efforts. On the other hand, the 'unplanned' *garimpo* towns, although having in some cases reached some degree of organised provision of services for the needs of their populations, have less pronounced urban characteristics than the older centres and struggle with a lack of sustained investment. Moreover, increasing control over the *garimpo* is starting to affect their migration to the mine sites. Current artisanal diamond mining, as opposed to wartime *garimpo*, often takes the form of small pockets of miners, operating clandestinely in small groups or even individually, in hidden areas. Van Lohuizen (2005, 3) describes how 'thousands of people dig up the earth in this lunar landscape', each mine employing between 10 and 50 workers spread over a vast terrain rather than working in urban concentrations. With the urban growth following the end of war, the trend is for informal mining towns to fade away.

Formal mining, together with government preferential investments in services in major established urban centres, is foundational to the distinction between planned and unplanned urban areas. Indirectly, informal mining earnings have always been channelled to the planned cities. Broadly, during the civil war the profits from the arms trade for diamonds circulated through national and international networks and, except for the arms, little or nothing was invested in the development of the mining towns. On the ground, illegal earnings were more easily detected if spent in the informal areas. The uncertainties of war largely prevented local investments, except for precarious, temporary trading posts, shops or canteens. By the end of the war, with the return of formal mining, and the continuation of *garimpo*, income generated by diamonds, through formal salaries and informal earnings was still being invested and spent primarily in the planned cities. Thus a selective urban development bias was mapped onto the country's inherited colonial geography, reflecting the highly uneven spatial accumulation of capital referred to by David Harvey (2005) and Neil Smith (2008).

The preference for investment in the planned cities relates not only to efforts to deflect detection of illegal earnings by the authorities. Other conditions are increasingly more decisive; the most important being that people find better infrastructure and access to services there. Moreover, the attractive pull of city life is irreversibly present in the imaginary of Angola's wartime migrants. In the *garimpo* mining towns, many shops and businesses are now closed whereas in the larger cities, new businesses, markets and shops are appearing at the same time as more houses and neighbourhoods are being built, extending the city's limits beyond the colonial and wartime boundaries.

Better off in the city: poverty and welfare in the twenty-first century

Three main conditions are currently contributing to improved living conditions in Angola's major planned urban centres, albeit unevenly. First, government investments in infrastructure are preferentially made in the administrative and economic centres of the older cities. Second, the government's increasing control over informal mining discourages settlement in old wartime unplanned urban settlements and any new *garimpo* towns. *Garimpo* profits are now less likely to be channelled into the growth of the unplanned towns, while incomes from formal mining jobs are being spent in the cities. Finally, the Lundas' recovery from war through the re-establishment and expansion of mining corporation investment is driving a new phase in the urbanisation of the provinces.

Despite its mineral wealth and rapid economic growth rate, Angolan poverty indicators are invariably high. In 2000, Angola's population in extreme poverty (earning less than $1.25 per day) comprised 54% of the total, and population in absolute poverty (living below $2 per day) was 77% (UNDP 2010). These figures decreased to 26% and 38% respectively in 2010. Yet the World Bank classifies Angola as a lower-middle-income country, and it may soon be categorised as a middle-income country in terms of its average GDP per capita. Extremely unequal income distribution accounts for the anomaly.

It is estimated that 37% of the population lives below the national poverty line (INE 2010). The lowest quintile of the population earns on average less than US$15 a month. The urban population is considerably better off generally than rural dwellers: 58% of rural Angolans are below the national poverty line as opposed to 19% in urban areas. Nonetheless, INE figures indicate that 79% of the urban population live in houses built with non-appropriate materials and 43% reside in overcrowded houses.

Poverty is, therefore, a major issue affecting urban development and welfare in Angola and the Lundas in particular. Government investment in education, health and other social services is visibly improving, albeit very gradually. There is evidence that the number of classrooms, enrolled students, hospitals and health centres in the Lundas provinces have registered an increase (GPLN 2008; GPLS 2008). State and administration reforms mean more jobs in the administration, education or health areas, which attract more people and divert a considerable number of youths from the *garimpo*.

In addition to government and mining corporate investment in the planned cities, these cities also have become the quasi-exclusive recipients of formal mining incomes and *garimpo* profits. Diamond-related profits and income reach the cities through individual investments in housing, commerce or services. Money earned both in the formal economy and by *garimpo* is spent and invested in the larger cities where the returns to investment are considered to be higher. This gives even more momentum to the growth of the planned cities like Saurimo and Dundo, while simultaneously attracting more people from the rural areas to the cities.

The new policies and legislation did not affect the status of *garimpeiros* after the end of the war. They were illegal under the old regulations, and remained so (Smillie 2009). But as artisanal mining is difficult to control in this region, new legislation of 2011 has legalised these activities with the aim of keeping them under administrative surveillance. Poverty associated with precarious living conditions in the declining informal 'unplanned' mining sites is a reality, contrasting starkly with the far better living conditions, infrastructure and services of urban areas.

On the other hand, the government is now more active in encouraging legitimate local investment of diamond profits. Since the cessation of war, the government has concentrated on regularising the ways and means of channelling diamond revenues to the Lundas. As van Lohuizen (2005, 1) notes: 'working conditions are still shockingly bad. Despite enormous profits, little of this income reaches the population'. The government launched a stabilisation programme for the diamond sector (PROESDA – *Programa de Estabilização do Sector dos Diamantes*) in 1997 and in 2004 Endiama established the Fundação Brilhante as the 'social arm' of the company, with the purpose of implementing cultural and social community development projects. However, the fund's actual performance has been heavily criticised (Gordon 2005; Bermúdez-Lugo 2010; USAID 2008).

There is now more pressure on mining companies to invest locally. The post-war role of the mining companies stretches beyond economic and industrial management to the provision of social services for the wider population. 'Even today, there is a perception that the mining companies, rather than the state, are responsible for the welfare of the population' (Pearce 2004, 58). The patronage role of the colonial mining companies has been replaced by what has been described as a 'state-like presence': catering for a population that work for, are surrounded by, or simply engulfed by mining concessions, corporate authorities become a primary institutional interface through which the population accesses consumer goods and services, as well as education and medical care (Calvão 2011).

The revision of national mining policies is geared towards returning 10% of diamond taxes to the Lundas. Mining companies are expected to invest in local development including infrastructure, schools, agriculture and medical posts, improving local facilities for workers and their dependants. Many of the Lundas operating companies have developed concrete plans and publicise them, showing their commitment to the provision of social services. The 2005 Annual Review of mining in Angola highlights the achievements, including taxes paid and investments in education, health, agriculture, electricity and water provision of Catoca and Sociedade de Desenvolvimento Mineiro (SDM), the two major companies operating in the region, that account for 56% of the country's total diamond production (Gordon 2005, 7–8). Nevertheless, despite company websites publicising mining investments in social areas, observers are still sceptical about the mining companies' investments and services to the workers and local population, compared to their profits.

The importance of the diamond sector to the region, whether formal or informal, is very visible. The negative effects of the international crisis of 2008 caused not only a slowdown of the economy in the region but, combined with government control, also made unplanned mining towns unattractive. Smillie (2009, 3) notes: '[diamond mines] have shut down or slashed production, exploration has come to a dead halt, and for the first time in nearly a decade, Angola no longer had foreign suitors lined up begging for a diamond concession'. Estimated diamond carat production declined (Coakley 2003; Bermúdez-Lugo 2010). This downward trend reduced the Lundas population's income-earning in mining, causing many to search for alternative sources of work, mostly in the major planned cities where government investment is higher.

There is a very long road ahead to address the inherited yawning divides between Angola's rich and poor, formal and informal mining and rural and urban areas. Until these gaps are narrowed, Angolans are likely to continue to migrate to the cities and *garimpo* will persist as a significant source of income for the poor.

Conclusion

Angola's long urban history has experienced a dramatic escalation of urbanisation in recent decades, against the backdrop of a civil war that drove the rural population to the cities. The Lundas context is particularly striking with respect to the agglomerations of diverse populations that emerged during that period. The impact of mining on urban growth first surfaced during the late colonial period. Mining company investment, coupled with national and regional urban settlement and management policies, led to the creation of structured and planned mining towns. In the Lundas, urban centres associated with dynamic companies and high levels of mine production outgrew other towns that were formally established to administer the territory. This is illustrated by Dundo, which gained economic, demographic and even administrative primacy over Lucapa, the official capital of the province.

Military control, population movements and relocation of mining sites during the civil war that followed deeply disrupted the planned towns. There was an exodus of most mining companies from the war zone areas. However, some of the mature older cities experienced urban growth caused by the population fleeing war conflict in the countryside.

The destiny of small-scale Angolan mining sites has been intimately connected with a torturous civil war. Local populations in Lundas sometimes had to resort to working at specific locations where the guerrilla and other illegal groups carried out mining activities because of an almost complete lack of opportunities elsewhere. The influx of people from neighbouring countries coming to dig for diamonds and search for business opportunities accelerated the growth of these sites and new towns emerged without any semblance of planning.

Over the last decade of peace, a few of these new towns have continued to accommodate large populations. Others, however, have become unattractive due to combined government control of the *garimpo* and increased investment in already planned cities and towns. Urban growth rates are fluctuating with the flux in mining in both formal and informal mining settlements. However, earnings from employment in the formal sector and from *garimpo* are no longer being channelled to spontaneous mining towns. Individuals prefer to invest in planned cities where living conditions are better. The urban agglomeration bias of capital investment (Harvey 2005; Smith 2008) leads to the retention of an urban primacy trajectory that keeps the mineral-rich periphery dependent on the metropolitan and regional centres, with adverse consequences for the national distribution of wealth. Small-scale mining implies far more makeshift and fluctuating patterns of settlement, which testify to the extensive poverty in the country and the experience of 'war urbanisation'. Given Angola's long-established urban planning legacy, the informal mining settlements stand out as far more spontaneous and indeed precarious towns compared to towns with a history of fostered and planned growth. Due to their colonial origin and superior infrastructure, the large mining company-related towns in Lunda attract the most investment, both from state and private companies and also from individuals. These cities, despite the vagaries of the civil war and massive informal settlement during the period, have retained their basic urban structure and administrative importance.

Informal cities, on the other hand, which have arisen from unplanned settlements, are mostly ignored as targets of business and infrastructural investment. Cuango is probably the only exception to this pattern, demonstrating that in rare cases, informal settlements may give rise to the emergence of a larger city capable of attracting government investment.

Awareness of the interplay of mining and urbanisation is essential to understanding Lundas' current development and its future possibilities for expansion of services and commerce to augment the core mining activities of the provinces. Likewise, these dynamics need to be considered in relation to the fluctuation of international diamond prices. Despite current government and mine companies' efforts to plan and provision urban development, the diamond price will always remain a wild card to contend with in the planning process.

Notes

1. União Nacional para a Independência Total de Angola.
2. Three years after independence, colonial Lunda district was divided into two administrative provinces, Lunda Norte and Lunda Sul. Henceforth, in this analysis 'Lundas' refers to these two provinces.
3. *Garimpo* is translated as 'mining' or 'digging'. The Portuguese term associates this activity with eighteenth century explorers in Brazil and the rudimentary equipment used then and now. It appears in Portuguese dictionaries at least since 1877 as a place where precious minerals like diamonds and gold are exploited. The term is also used to designate the activity of *garimpeiros*, known elsewhere as artisanal or small-scale miners.
4. According to the latest estimates, about 5 million of Angola's 18 million national population are resident in Luanda (UNDP and UNFPA statistics).
5. The *Planos de Zona de Ocupação Imediata* (PZOI) (immediate occupation zonal plans), Angola's first urban plans were launched in 1973 and 1974 in some of the larger Lunda cities and towns: Luachimo, Cambulo and Portugália associated with Dundo; Xa-Muteba, Cuango and Saurimo (Fonte 2007).
6. The district of Lunda was created on 13 June 1895. Malanje remained the capital for several years.
7. The pre-colonial history of the Lundas hints at the mineral wealth of the region. Blacksmiths were important in Lunda and nearby Kongo, which stimulated early prospecting (Dias 2003).
8. In 1906, diamonds had already been found in the Chikapa river basin.
9. Translated as 'Angolan Mining Research Company'.
10. The founding members of PEMA were the Banco Nacional Ultramarino and Henry Burnay & Co. (later Banco Burnay) in Portugal; Société Génerale de Belgique and Mutualité Coloniale in Belgium; Banque de L'Union Parisienne in France; and the Ryan-Guggennheim group in the United States.
11. The members of the Companhia de Diamantes de Angola – Diamang derived from PEMA and the Anglo-American Corporation of South Africa. The company's articles of association were published in 1918 (*Diário do Governo*, II Série, n° 136 of 12 June and *Boletim Oficial da Província de Angola*, II série, n° 27 of 6 July).
12. Kimberlite diamonds are the most profitable for large-scale mining and to date more than 600 such pipes have been discovered in Angola.
13. These workers would usually come to work in the mines on a one-year compulsory contract, but many would remain in the diamond villages afterwards. Others returned to their hometowns with considerable knowledge of the diamond business.
14. For instance, after the 1940 rebellion, successive waves of Cuvale prisoners were sent to the mines, and many other workers were taken when they were captured by the administrative authorities for default on taxes.
15. Diamang was the gravitational centre of Dundo. It became one of the largest employers in Angola and this forced the company to establish rules on workers' hygiene and health and to build infrastructural support. The city was organised around the compound, with socially and economically stratified encircling areas radiating from the centre.
16. Diamang's pioneer diamond project, the Luxinge project in Nzagi (Cambulo), 85 km from Dundo.
17. With paved streets, a number of high-rise buildings, houses, stores and warehouses, Saurimo became a city on 28 May 1956 (Decree 2757). Its Cokwe name means *Sa* (owner) and *limbo* (land).

18. Owned by Sociedade Mineira de Catoca Lda., 35 km from the city, it is the fourth largest diamond mine in the world. The Catoca kimberlite pipe in Angola was discovered in 1985 but the civil war in Angola hampered its early development, which finally began in 1997 (Read and Janse 2009, 2).

19. Endiama – *Empresa Nacional de Diamantes de Angola*, EP is a state-owned company set up in 1981 (Decree 8/81) by a committee of the National Defence and Security Council to operate the diamond mining concession.

20. *Movimento Popular de Libertação de Angola*, People's Movement for the Liberation of Angola.

21. 'Blood diamonds' (*diamantes de sangue*) or 'conflict diamonds' are diamonds from geographic areas controlled by forces or factions opposed to internationally recognised governments and which are used to finance military action against those governments and do not comply with the decisions of the UN Security Council. In December 2000, the General Assembly of the UN adopted a resolution against the role of diamonds in financing military conflicts. Angola and Sierra Leone were specifically named by the General Assembly. Since 1998, there had been sanctions against UNITA. The institution of the Kimberley Process that followed contributed to stricter control mechanisms for the extraction, certification and export of diamonds globally and particularly in Angola. It gradually set the stage for regaining control over the mining sites and cities associated with them. The Kimberley Process Certificate Scheme (KPCS) is a late 2002 industry-wide effort to prevent trading in rough diamonds by insurgent groups, involving 75 countries. The current certification system is controlled by SODIAM and ASCORP in Angola (Marques 2005, 11).

22. USAID (2008, 17) figures say 150,000 to 300,000.

23. These companies must be authorised by the Ministry of Geology and Mines and the public diamond company Endiama.

Note on contributors

Cristina Udelsmann Rodrigues is an anthropologist specialised in African studies, working at ISCTE-IUL, University Institute of Lisbon as an Associate Researcher. Her main research areas include urban anthropology, poverty, borders and urban strategies. Her urban publications include: From family solidarity to social classes: Urban stratification in Angola (Luanda and Ondjiva). *Journal of Southern African Studies* 33, no. 2 (2007): 235–50; Angolan cities: Urban (re)segregation?, in F. Locatelli and P. Nugent (eds) *African cities: Competing claims on urban spaces* (Leiden: Brill, 2009), 37–53; Angola's southern border: Entrepreneurship opportunities and the state in Cunene. *Journal of Modern African Studies* 48, no. 3 (2010): 461–84. She can be contacted at: crisrodrigues70@gmail.com

Ana Paula Tavares is an historian who has specialised in the areas of culture, museums, patrimony and ethnology. She has worked as a cultural delegate in Kwanza Norte and at the Angolan Historical Archive and the Angolan National Centre of Documentation and Historical Research. Her main research area is Angolan history, particularly the Lundas region. Previous publications on this subject include *Na mussumba do Muatiânvua quando a Lunda não era leste. Estudo sobre a Descripção da Viagem à Mussumba do Muatiânvua de Henrique de Carvalho* (Lisbon: University of Lisbon, 1995); As Lundas (2010), in *Rumo às Terras que Brilham: Lundas* (Lisbon and Luanda: Pangeia and Chá de Caxinde). She can be contacted at: apaulatavares@yahoo.com

References

ADRA. 2004. Aldeias da Lunda-Sul. *Estudo de Caso* 5. ADRA for the Rede Terra.

Allen, M. 2010. Cafunfo, Angola. *The Wall Street Journal*, June 19. http://online.wsj.com/article/SB10001424052748704198004575311282588959188.html.

Amaral, I. 1962. Ensaio de um estudo geográfico da rede urbana de Angola. *Estudos, ensaios e documentos*. Junta de Investigação do Ultramar, 97.

MINING AND AFRICAN URBANISATION

Bermúdez-Lugo, O. 2010. The mineral industry of Angola. *2008 Minerals Yearbook*. Reston, VA: US Department of the Interior and US Geological Survey.

Bryceson, D. 2006. Fragile cities: Fundamentals of urban life in East and Southern Africa. In *African urban economies: Viability, vitality or vitiation?*, ed. D.F. Bryceson and D. Potts, 2–38. London: Palgrave Macmillan.

Calvão, F. 2011. When boom goes bust: Ruins, crisis and security in megaengineering diamond mining in Angola. In *Engineering earth*, ed. S. Brunn, 367–82. Heidelberg, Germany: Springer.

Cleveland, T. 2008a. Rock solid: African laborers on the diamond mines of the Companhia de Diamantes de Angola (Diamang), 1917–75. PhD diss., University of Minnesota.

Cleveland, T. 2008b. *Life and labor on the diamond mines of the Companhia de Diamantes de Angola (Diamang), 1917–1975*. University of Minnesota. http://www.ces.columbia.edu/pub/papers/Cleveland.pdf.

Coakley, G.J. 2003. The mineral industry of Angola. *Minerals Yearbook 2003*. Reston, VA: US Geological Survey.

Costa, A., and C.U. Rodrigues. 2008. Famílias e estratégias de sobrevivência e reprodução social em Luanda e Maputo. In *Subúrbios de Luanda e Maputo*, ed. I. Raposo and J. Oppenheimer, 139–61. Lisbon: Colibri.

De Boeck, P. 2001. Garimpeiro worlds: Digging, dying and 'hunting' for diamonds in Angola. *Review of African Political Economy* 28, no. 90: 549–62.

De Carvalho, H.A.D. 1890–1894. *Descripção da viagem à Mussumba do Muatiânvua*. 4 vols. Lisboa: Imprensa Nacional.

Dias, J. 1989. Relações económicas e de poder no interior de Luanda, ca. 1850–1875. *1ª Reunião Internacional de História de África: Relações Europa-África, no 3° quartel do século XIX*: 241–58. Lisbon: Instituto de Investigação Científica Tropical, CEHCA.

Dias, J. 2003. Caçadores, artesãos, comerciantes, guerreiros: Os Cokwe em perspectiva histórica. In *A antropologia dos Tshokwe e Povos Aparentados*, ed. L. de Heusch and J. Dias, 17–47. Porto: Faculdade de Letras da Universidade do Porto.

Dietrich, C. 2000. Porous borders and diamonds. In *Angola's war economy*, ed. J. Cilliers and C. Dietrich, 317–45. Pretoria: Institute for Security Studies.

E-Polis. 2010. *Urban populations 1880–2020*. http://egeopolis.eu/spip.php?article193&id_article=193; http://e-geopolis.eu/IMG/png/AFCE_SEMI_2010.png.

Ferrão, C.A. 2005. *Minas e pantominas: Aspectos positivos e negativos da mineração*. Luanda: Novo Imbondeiro.

Fonte, M. 2007. Urbanismo e arquitectura em Angola: de Norton de Matos à revolução. PhD diss., Technical University of Lisbon, Faculty of Architecture.

Gordon, C., ed. 2005. *Diamond industry annual review: Republic of Angola 2005*. Ottawa: Partnership Africa Canada.

GPLN (Government of the Province of Lunda Sul). 2008. *Programa para o ano 2009*. Lucapa, Lunda Norte: GPLN.

GPLS (Government of the Province of Lunda Norte). 2008. *Programa para o ano 2009*. Saurimo, Lunda Sul: GPLS.

INE (Instituto Nacional de Estatísitica). 2010. *Inquérito integrado sobre o bem-estar da população (IBEP) 2008–09*. Principais Resultados Definitivos, Versão Resumida. Luanda: INE.

Isaksen, J., I. Tvedten, and P. Ilinga. 2006. *Experience and institutional capacity for poverty and income distribution analysis in Angola*. CMI, Report 19.

Harvey, D. 2005. *Spaces of neoliberalization: Towards a theory of uneven geographical development*. London and New York: Franz Steiner Verlag.

Jenkins, P., P. Robson, and A. Cain. 2002. Local responses to globalization and peripheralization in Luanda, Angola. *Environment and Urbanization* 14, no. 1: 115–27.

Le Billon, P. 2008. Diamond wars? Conflict diamonds and geographies of resource wars. *Annals of the Association of American Geographers* 98, no. 2: 345–72.

Malaquias, A. 2001. Making war and lots of money: The political economy of protracted conflict in Angola. *Review of African Political Economy* 28, no. 90: 521–36.

Malaquias, A. 2007. *Rebels and robbers: Violence in post-colonial Angola*. Uppsala: Nordiska Afrikaintitutet.

Marques, R., and R.F. de Campos. 2005. *Lundas: As pedras da morte, relatório sobre os direitos humanos*. Lisbon: Grafispaço.

Pearce, J. 2004. War, peace and diamonds in Angola: Popular perceptions of the diamond industry in the Lundas. *African Security Review* 13, no. 2: 51–64.

Pereira, E., J. Rodrigues, and B. Reis. 2003. Synopsis of Lunda geology, NE Angola: Implications for diamond exploration. *Comunicações do Instituto Geológico e Mineiro*, no. 90: 189–212.

Potts, D. 1997. Urban lives: Adopting new strategies and adapting rural links. In *The urban challenge in Africa*, ed. C. Rakodi, 447–94. Geneva: United Nations University Press.

Read, G.H., and A.J.A. Janse. 2009. Diamonds: Exploration, mines and marketing. *Lithos* 112S: 1–9.

Rodrigues, C.U. 2007. From family solidarity to social classes: Urban stratification in Angola (Luanda and Ondjiva). *Journal of Southern African Studies* 33, no. 2: 235–50.

Rodrigues, C.U. 2009. Angolan cities: Urban (re)segregation? In *African cities: Competing claims on urban spaces*, ed. F. Locatelli and P. Nugent, 37–53. Leiden and Boston: Brill.

Simone, A., and A. Abouhani, eds. 2005. *Urban Africa: Changing contours of survival in the city*. Dakar: CODESRIA.

Smillie, I., ed. 2009. Diamonds and human security. *Annual review 2009*. Partnership Africa Canada. http://www.pacweb.org/Documents/annual-reviews-diamonds/AR_diamonds_2009_eng.pdf.

Smith, N. 2008. *Uneven development: nature, capital and the production of space*. 3rd ed. Athens, GA and London: The University of Georgia Press.

Tavares, A.P. 2010. As Lundas. In *Rumo às terras que Brilham: Lundas*, ed. M. Anacoreta Correia, 69–80. Lisbon and Luanda: Pangeia and Chá de Caxinde.

United Nations Cartographic Section. 2008, Map no. 3727 Rev. 4, August. http://www.un.org/depts/Cartographic/english/about.htm.

UNDP (United Nations Development Program). 2010. Human development index. http://hdrstats.undp.org/en/countries/profiles/AGO.html.

USAID. 2008. *Avaliação de viabilidade e diagnóstico de desenvolvimento de Cuango*. Washington: MSI.

Van Lohuizen, K. 2005. Diamond matters: From the mines to the jet-set. http://www.cred.org.uk/data/credfoundation/downloads/resources/DMhandleiding.pdf.

Vlassenroot, K., and K. Büscher. 2009. The city as frontier: Urban development and identity processes in Goma. *Crisis States Working Papers Series*, no. 2. London School of Economics, Crisis States Research Centre.

Diamond mining, urbanisation and social transformation in Sierra Leone

Roy Maconachie

This contribution critically explores changing relationships between diamond mining and patterns of urbanisation in Sierra Leone. In providing an historical overview of mining expansion and contraction, the paper highlights the significant impacts that mining has had on the rural–urban continuum, and how this has shaped political, economic and social change in diamondiferous regions. Focusing on Kono District, the effects of diamond mining on populations are evaluated before, during and after the civil war, demonstrating how diamonds have had diverse and varying impacts on both population mobility and urban agglomeration at different points in time. While much attention has focused on the social consequences of wartime displacement from diamondiferous areas to the capital city, Freetown, recent research suggests that the postwar return of young people to diamond mining regions has had unexpected consequences. Most significantly, a decline in artisanal mining activities and the rise of large-scale industrial mining has reawakened the interest of young ex-miners in farming, especially those who enjoy hereditary land rights. While one consequence of the war may be that the population is now more urban and more mobile, the paper concludes that the return of young people to their villages of origin, and rapprochement with local chiefs, may be helping to drive a resurgence of community-based cooperation in Kono District, a development which could provide a more durable basis for sustainable and democratic development in the years to come.

Since their discovery in the 1930s, diamonds and the attendant seasonal migration of labour to diamondiferous areas have long caused significant social, political and economic transformation in Sierra Leone. Unlike countries in Southern Africa, such as South Africa or Botswana where diamond mining is associated with the mechanical mining of deep reserves, in Sierra Leone, diamond resources are more commonly dispersed in the gravels of river beds and terraces, are located close to the surface, and are available to individual diggers with little more than shovels, sieves and a source of water for straining the gravel. Indeed, the accessibility of alluvial diamond deposits and low barriers to entry for artisanal miners have always been a magnet for migrant 'strangers' to Kono District, mostly young, single, uneducated, unemployed men seeking to make their fortunes. A recent study in Sierra Leone's diamondiferous areas confirms the continuing importance of population mobility in these regions, suggesting that some 10,000 seasonally mobile workers support between 70,000 and 140,000 poor people in sending districts (Amco-Robertson Mineral Services 2002).

While low barriers to entry have made it possible for many households to draw upon artisanal mining as an important strategy in their livelihood portfolios, the ease with which Sierra Leone's alluvial diamonds can be mined and exported has also undermined the state's ability to direct diamond revenues through official channels and harness their benefits for development (Brown et al. 2005). While it is both easy and economically practical for mining companies to strictly monitor deep pit mechanical mining, it is virtually impossible to closely control alluvial diamond fields that are mined artisanally. In Sierra Leone, two river systems, the Sewa and its tributaries, flowing through Kenema, Bo and Bonthe Districts, and the Moa, flowing through Kenema and Pujehun Districts, have deposited diamonds over large areas in

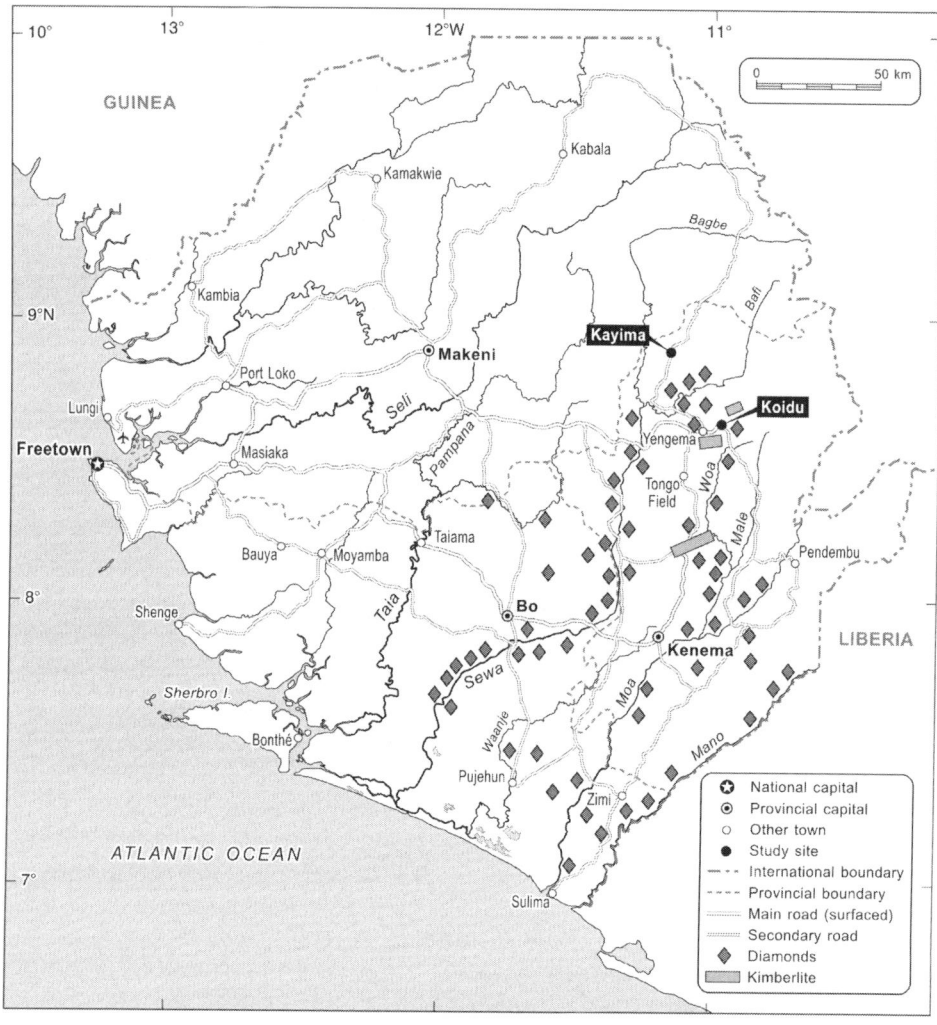

Figure 1. Diamondiferous regions of Sierra Leone.
Source: Author's fieldwork.

the south and east of Sierra Leone (Figure 1). The country's alluvial diamond mining fields cover an area of almost 20,000 km^2, although the actual diamond-bearing alluvial ground is only about 200 km^2. Central administrative control over artisanal mining is thus extremely difficult and has been a significant factor in the historically high rates of diamond smuggling from Sierra Leone. Recent estimates of the value of diamonds lost to smuggling range from 50% to as much as 90% of total production by value (Partnership Africa Canada 2004).

While artisanal diamond mining is commonly undertaken over a vast expanse of the country's Eastern Province, and the highly localised 'point development' often associated with the exploitation of other mineral resources has not transpired, diamonds have shaped wider development processes and outcomes in a variety of other ways. For example, diamond mining activities have had diverse and varying impacts on population mobility and patterns of urban agglomeration at different points in the country's colonial and post-colonial history. In the past, both formal company mining and informal indigenous extraction have stimulated temporary and permanent migration, the emergence of new towns and the expansion of existing ones, and a strong element of seasonality with populations moving between mining and farming activities over the course of the year (Maconachie and Binns 2007; Swindell 1974). More recently, since the end of the country's civil war in 2002, a steady increase in urban populations, and particularly an influx of young people into the diamond mining areas, has had significant social, political and economic consequences in Kono District. Population growth rates in diamond mining areas continue to rise, with reports from UNAMSIL suggesting that the population of Kono District has grown annually at a rate of nearly 150,000 persons over the past several years (Brown et al. 2005).

This contribution explores the relationship between diamond mining, social transformation and changing patterns of settlement in Sierra Leone, by presenting an historical overview of mining expansion and contraction and its impact on the rural–urban continuum. In particular, three main historical periods of time are critically explored, demonstrating how a wide range of factors at different scales have shaped mining settlements over time. The first section focuses on the colonial years, beginning with the discovery of diamonds in the 1930s and extending until the time of independence in 1961. During this period, the first major 'diamond rush' to Kono District took place in the early 1950s, an event that fuelled a process of regional urbanisation to mining towns. This movement of people had both social and political-economic ramifications, prompting fears from colonial officials that diamond mining was breaking the bonds of traditional communities and the authority of rural elders, as young able-bodied people were drawn away from farming. But through the 1950s and up until the time of independence, 'downstream' industries and periodic markets increasingly developed around mining settlements and the links between diamondiferous areas and regional cities and towns were strengthened, most notably the burgeoning Kono town of Koidu.

The second period of analysis in the paper spans from Sierra Leone's early years of independence – including the difficult period of rule under Siaka Stevens and the one-party state – up until the time of the civil war of the 1990s. During this period, the changing nature of the state and the entrenchment of patronage politics served to exacerbate the weakness of formal state institutions. Diamonds became a key strategic resource for Stevens' regime and the diamondiferous areas continued to attract considerable influxes of population. During the 1970s and 1980s in particular, urban populations in the country's diamondiferous regions mushroomed in size, generating various regional multiplier effects for the local economy. On the one hand, diamonds served as an impetus for local economic development as the increasing demand for foodstuffs in urban centres and mining areas continued to stimulate the development of periodic markets and the growth of a large and varied group of itinerant traders. But on the other hand, as diamonds became a strategic political tool for Siaka Stevens' All People's Congress government, the diamond sector became a sphere of unregulated private enterprise, eventually leading to state collapse and mass internal population displacement. Most notably, the impact of forced displacement from Kono District during the civil war years had severe ramifications, when rebel insurgents took control of Kono's diamond fields and local residents were forced to flee to the safety of Freetown, causing the population of the capital city to almost double in size.

Finally, in the third period of analysis, the impact of the civil war on diamond mining activities and the implications of population displacement from Kono District are discussed. Since the end of the war, as many internally displaced people have returned to their homes in Kono, both rural and urban populations have been steadily growing in the post-conflict period. However, recent fieldwork suggests that as many alluvial diamond areas have become 'worked out', mining activities have increasingly shifted from small-scale artisanal extraction to larger-scale mechanical mining, which has subsequently reawakened the interest of many young ex-miners in farming, especially those who enjoy hereditary land rights. While one consequence of the war may be that the population has now become more urban and more mobile, the return of young people to their villages of origin, and rapprochement with local chiefs, may be helping to drive a resurgence of community-based cooperation in rural Kono District, a development which could provide a more durable basis for sustainable and democratic development in the years to come.

Settlement growth, urbanisation and mining

Colonial years, 1930–1961

Diamonds were first discovered in the Gbobora stream in Kono District in the Eastern Province in 1930, and further geological exploration subsequently revealed a large alluvial diamond source area in the central plains of Kono District. When mining operations first began in 1932, the colonial government's original plan for tapping the revenue-generating potential of diamond mining was to grant an exclusive licence to explore, mine and market diamonds to the Sierra Leone Selection Trust (SLST), a wholly owned subsidiary of Consolidated African Selection Trust. In 1934, when the SLST began operations, it thus obtained a monopoly over the production and marketing of all Sierra Leonean diamonds, excluding only the areas already designated for iron ore mining (Saylor 1967). In 1953, a second major source area was discovered in Lower Bambara Chiefdom in Kenema District, where diamonds found in the basin of the Male River were traced via the Woa River to the Tongo River (Binns 1982).

Colonial reports suggest that during this period, a considerable amount of diamond revenue was being reinvested in local development (Temple 2006). At the time, employees of the SLST were not only provided with relatively well-paid jobs, but they also received a range of associated benefits, including schooling for their children, university scholarships, housing and medical care. Reports also indicate that during these early days, diamond revenues were used to finance infrastructure development, including road construction, and the provision of clean drinking water and electricity (Temple 2006). Consequently, in many respects, the settlements in diamondiferous areas – most notably in Kono and Kenema Districts – prospered and were important drivers of economic growth.

By the late 1940s and early 1950s, stories of the diamond frenzy had spread. Large numbers of people, particularly young males, began arriving from both within and outside Sierra Leone into the mining areas with the hope of recovering a share of the diamond wealth. Many were encouraged by the arrival of numerous foreign supporters and dealers, principally from Guinea, and illegal mining operations became widespread in and around the SLST reserves. By 1954, illicit mining operations had gradually spread along the Sewa and Bafi rivers southwards as far as Bo and Kenema Districts (Van der Laan 1965).

This mass influx of population into the diamondiferous regions had significant but varying impacts on the local community. On one hand, there was an immediate and well-documented decline in the production of foodstuffs throughout large parts of the Eastern Province. During this time, imports of rice and meat were necessary to alleviate food shortages, and export crop production declined (Gberie 2005). In urban settlements – predominantly the towns of Kenema and Koidu, the two dominant urban centres in the mining areas during the late 1950s and early 1960s – food initially had to be imported from abroad and from areas of surplus production. However, as the patterns and processes associated with the coexistence of mining and agriculture became more established, other trends emerged. For example, evidence suggests that diamond winnings were increasingly being converted into cash and consumer goods were sent back to the rural areas (Binns 1982). As a result, the links between farmers and miners grew and many farmers became involved in mining activities in addition to their farming interests. Thus certain communities greatly benefited from the achievements of the mining population. Furthermore, despite the unpopularity of the destruction of good farm land in the mining areas, farmers increasingly realised that the growing urban centres, with their mainly non-farm populations, were attractive markets for foodstuffs and significant incomes could be generated from sales in these areas.

As such, historically, the growth of settlements in the diamondiferous regions has not simply been the product of an influx of miners to diamond-bearing regions, but has also reflected a major population expansion associated with a series of 'downstream' industries based on the need for food, shops, traders and other services to cater to the mining population. This was certainly the case in the burgeoning towns of Kenema and Koidu during the late 1950s and early 1960s. In the case of Kenema, Gamble (1964) estimates that in 1909, the settlement only had about 30 houses but by 1946, it had grown to a town of some 4000 residents. By 1963, the population had expanded to 13,000 people, and by 1974 this had increased to an estimated 31,300, representing a growth rate over the 11 year period of over 136%, making it the fourth largest town in the country at the time. As Gamble (1964) recounts, this process of urban growth had many positive multiplier effects on the local economy:

The rush to the diamond areas in the middle 1950s, brought many strangers and led to a great increase in the population. Some of the newly made wealth was invested in houses, many Mende preferring to build in town rather than in their home villages, and in lorries and buses, leading to the expansion of the town as a commercial centre for the surrounding region. (Gamble 1964, cited in Binns 1982, 25)

While the expansion of Kenema during the early years of the mining boom was impressive, the most significant example of urban growth in Kono District during the 1950s and early 1960s was undoubtedly that of the town of Koidu. In 1928, before the discovery of diamonds, Koidu was a settlement of only six huts. However, the diamond rush of the early 1950s resulted in a vast influx of migrants from both within the country and the wider sub-region, converging on areas around Koidu where diamonds had been discovered. This resulted in many people giving up their traditional work and moving into mining as a full-time occupation, and by 1963 the town's population had grown to 14,309 people. The initial transformation of Koidu is described by Binns (1982, 25):

> Until the early 1930s when mining started, the area was entirely devoted to farming. No major trade routes passed through Koidu, and the main line of transport, the railway completed in 1908, ran 60 miles to the south. Koidu developed near to the SLST number 2/4 washing plant, about one mile south-east of Sefadu, the District headquarters. The main street of the town was in fact the dumper line, along which trucks passed taking gravel to the plant. In the 1950s and early 1960s, Koidu's growth was very rapid as the town became the main market centre for the area.

While the physical transformation of settlements such as Koidu from villages to bustling urban centres was an astonishing consequence of the discovery of diamonds, there were also significant social and political knock-on effects associated with mass engagement in artisanal mining in a predominantly agrarian economy administered at the local level by customary authorities. While the Sierra Leone Protectorate, attached to the original coastal colony in 1896, was always administered under a system of 'indirect rule', efforts to strengthen the agrarian base of the Protectorate economy in the early twentieth century saw chiefs being granted considerable powers over their subjects. In Kono District, these powers included the right to tribute and to punish subjects for travelling outside their chiefdom without permission and for failing to engage in productive labour. While the heavy fines that chiefs imposed on their subjects only served to suppress peasant economic enterprise, the British colonial authorities were always inclined to overlook these shortcomings as long as they continued to maintain law and order. Chiefs served as the primary authorities for all African 'natives', and migrants into the mining areas were especially dependent on their patronage when seeking places to live and land to work.

State interventions to manage the diamond industry exposed the inherent weaknesses and contradictions of this system of local administration. Although the government-run SLST provided a legal basis for controlling illicit African mining, there were no uniformed police in the Protectorate at the time, and the colonial government authorised the SLST to deploy its own armed security force. In 1952, Kono became the first district in the Protectorate to be policed by the colony-based Sierra Leone Police Force. Police activity increased with the diamond rush, but in certain areas resistance was encountered from well-organised gangs of diamond diggers who had their own protection squads. Violent conflict between rival mining gangs frequently ensued and crimes committed in the diamond areas were increasingly reported in the national press (Binns 1982). As SLST security forces cracked down on illegal artisanal mining activities in their concession areas, chiefs viewed this as an infringement upon their authority, not least because they had been actively brokering the local settlement of increasing numbers of African migrants eager to gain access to the diamond fields and happy to pay them rents and tributes for that privilege. In an attempt to appease the situation, the SLST made private payoffs to local chiefs in a vain attempt to stem the influx of illicit miners (Reno 1995, 47–9).

British colonial officials were also concerned that the influx of population into the diamond mining areas might undermine customary authority and feared that the bonds of traditional communities and the authority of elders would be broken as young able-bodied people were drawn away from farming. For example, the annual report on provincial administration for 1953 observes that a 'large number of adventurers of undesirable type' were flocking to the diamond fields of Kono District to 'make fortunes which they dissipated as irresponsibly as they came by them' (Government of Sierra Leone 1954, 5). The corresponding report for 1955 warns that:

> the atmosphere prevalent in the diamond mining areas, which was spread by miners retuning from them with money to spend and dissolute habits, could not be other than inimical to the traditional restraints of tribal discipline and good order, and some apprehension was felt on this account. (Government of Sierra Leone 1956, 4)

Aggressive policing of the SLST concession area around Koidu tended to exacerbate local grievances and led many Kono residents to believe that they were not getting a fair share of the wealth generated from their land. This quickly raised concerns about the ever-increasing influx of migrant 'strangers' into Kono District, especially wealthy members of ethnic trading diasporas (Lebanese, Fula and Mandingo) who were supporting illicit mining but were perceived to be reinvesting the profits elsewhere (Hayward 1972).

Measures to check immigration into Kono District and slow down the rapidly expanding population in diamondiferous areas were taken in 1954 and were directed initially against the Lebanese. A bill was subsequently passed in May 1955 to restrict the entry of black African immigrants into mining areas, most notably Mandingos from Guinea. On a visit to Kono District in October 1956, the governor issued an ultimatum stating that foreigners should leave Sierra Leone within three weeks (Binns 1982). The expulsion order was reported as being successful beyond expectation, mainly because of the cooperation from neighbouring French authorities. When all police reports were finally compared, it was estimated that 45,000 black foreigners left Sierra Leone within three weeks of the ultimatum (Annual Report of the Colonial Office [United Kingdom Government, Sierra Leone 1956]). Nevertheless, according to official estimates, 75,000 people were still engaged in diamond mining in the entire country in the mid-1950s, most of them operating illegally (Gberie 2005).

In the meantime, efforts were made by the colonial authorities to legalise diamond digging for native Sierra Leoneans. The establishment of the Alluvial

Diamond Mining Scheme (ADMS) early in 1956 meant that indigenous Sierra Leoneans could apply for mining licences to legally extract diamonds in designated areas outside the two SLST mining company leases in Yengema (Kono District) and Tongo Field (Kenema District). Spurning an opportunity to bring modern bureaucracy closer to the people, the government left it to the chiefs to authorise private mining licences. However, in this respect, the ADMS only served to strengthen chiefs' brokerage role in the artisanal diamond industry. It also helped to entrench the 'tributor–supporter' system of mining governance in which investors ('supporters') would negotiate access to mining plots with chiefs and/or licence holders and hire youth gangs to work them. The mining gangs ('tributors') would be supplied with food and tools by the supporter but receive no other payments unless diamonds were found (Zack-Williams 1995). This system of mining has endured in Sierra Leone to this day, and because tributors are held in a form of 'debt bondage' to their supporters, some commentators have noted that for those at the bottom of the diamond chain, artisanal extraction in Sierra Leone resembles little more than a modern-day form of slavery (Smillie 2000; Zack-Williams 1995).

Independence and the patrimonial state, 1961–1991

When Sierra Leone gained independence in 1961, there continued to be much euphoria over the role that diamonds might assume as a driver of development, as both the local economy and size of mining settlements continued to expand. As Keen (2005, 12) notes, 'efforts were made to rationalise the economy, to bring production within the legal sphere, to increase the tax revenue from primary product exports, and to use the proceeds for developing the country's infrastructure and social services'. Although during the early years of independence there was significant progress in reining in illegal mining activities, generating substantial revenues from artisanal mining continued to prove a great challenge (Keen 2005). Government estimates suggest that in the early 1960s, diamond revenues produced under the ADMS were only one-thirtieth of their world market value (Van der Laan 1965, 27).

In an effort to gain more control over the industry, in 1970 the Sierra Leone Government acquired a 51% share in SLST and the National Diamond Mining Company (NDMC) was formed. At that time, extraction methods became more mechanised, involving draglines, earth-moving equipment and sophisticated treatment plants to separate diamonds from the gravels. Consequently, between 1963 and 1975, the sale of industrial and gem diamonds represented about 60% of the country's export revenue, with the total value of diamond production peaking at £19.3 million in 1975 (Binns 1982). However, illicit mining in the two NDMC leasehold areas (Yengema and Tongo Field) continued to thrive and frequently resulted in 'stranger drives' to purge illegal miners from settlements within close proximity to the lease areas.

During this period, migration to Koidu continued apace and by 1974 the town had grown to a population of 75,600 people. The physical size of Koidu also expanded considerably along its four major roads. Many improvements to urban infrastructure were funded by diamond revenues, including the rebuilding of the water supply and sanitation systems, many secondary roads and the central market (Binns 1982). The importance of the NDMC's road construction and maintenance opened up new areas for mining, while at the same time increasing mobility to many of the smaller towns of the central Eastern Province and creating new markets for goods and services. Thus, once again, as was apparent in the 1950s and early 1960s, continuing migration streams into the mining areas were being driven as much by multiplier effects and the arrival of young men and women seeking to provide services to the mining population as by the rush of miners themselves who were seeking to make their fortunes. Table 1 compares population census statistics over a 10-year period from the 1963 to 1974 census, and reveals that a significant proportion of towns with over 5000 people in 1974 were located in diamond mining areas. Most notably, the growth of towns such as Motema, Yengema and Tokpombu was particularly rapid.

One economic consequence of increased urbanisation in Kono District during the 1970s, particularly in some of the smaller settlements, was that many farmers living on the urban fringe adapted their traditional food production systems in order to supply more foodstuffs to the mining area markets. Within the mining areas, local markets grew rapidly and the links between towns and their food supply hinterlands were strengthened. According to Binns (1982, 5):

> It might be suggested that the diamond mining areas and their rural hinterlands have experienced a demand-led growth, where the large demand for foodstuffs from the non-producing mining population has had inflationary effects on the price of food. This has had detrimental effects on those people who buy their food rather than produce their own subsistence requirements. However, the increasing demand for foodstuffs has encouraged a large group of small farmers to move out of the predominantly

Table 1. Population of settlements over 5000 inhabitants – 1963–1974 compared.

Town	Population in 1963	Population in 1974	Growth rate 1963–1974 (%)
*Bo	26,613	39,400	48.0
*Kenema	13,246	31,300	136.3
*Koidu	11,706	75,600	428.3
*Yengema	7,313	14,600	99.6
*Yomandu	5,469	7,500	37.1
*Peyima	4,625	5,200	12.4
*Tokpombu	1,524	6,100	300.3
*Motema	1,124	5,500	389.3
Greater Freetown	127,917	274,00	114.20
Makeni	12,304	26,500	115.4
Lunsar	12,132	21,500	77.2
Magburka	6,371	10,400	63.2
Segbwema	6,258	6,900	10.3
Bonthe	6,230	6,900	10.8
Port Loko	5,809	10,500	80.7
Kailahun	5,419	7,200	32.8
Kabala	4,610	10,300	123.4
Moyamba	4,564	6,400	40.2
Rokupr	4,151	5,500	32.5
Kambia	3,700	5,700	54.0
Lungi	2,170	6,500	200.0

*Settlements in diamond mining areas.
Source: adapted from Binns (1982, 158).

subsistence mode of production and into the production of food for sale in the local economy.

Thus some commentators have suggested that the stereotypical view of the negative impacts of diamond mining on farming was in fact far from an accurate picture of reality. Rather, it was apparent that many part-time miners were also actively working and investing in both food and export crop production. For example, research undertaken in the 1970s by Binns in the town of Kayima, situated some 40 kilometres to the north-west of Koidu (Figure 1), illustrated that large quantities of food crops were being produced for sale to the mining population, and the income generated was being reinvested by producers into their homes, families and, most notably, in the expansion of cash crops such as coffee, cocoa and citrus fruits. Whilst many farmers were producing considerable amounts of surplus rice to sell in the mining area markets (Rosen 1974), the greatest response in terms of production system changes occurred with fruit and vegetable crops, which were increasingly grown for sale to the mining population.

According to Riddell (1974), unlike other parts of West Africa, periodic markets in Sierra Leone were relatively slow to develop, being stimulated into existence by growing urban centres associated with diamond mining. Most notably, the large fortnightly market at Tokpombu outside the NDMC mining camp in Tongo Field was initiated in 1970 by the local paramount chief to supply food and consumer goods to the growing mining population, and was commonly known as 'Pay Friday Market', since it coincided with the days when NDMC staff were paid (Binns 1982). Riddell (1974, 547–8) suggested that these periodic markets were both 'critical' and 'innovative', placing, 'a stimulus for commercial crop production near the farmer, for the first time in many areas, and they may commercialise an essentially subsistence system of agriculture'.

While the growth of Sierra Leone's diamond economy in the 1970s undoubtedly generated a significant number of regional economic multiplier effects in Kono District, particularly in its stimulation of small-scale agricultural activities, it also became evident that alluvial diamond mining was the type of activity that lent itself extraordinarily well to patrimonial politics (Richards 1996). Indeed, it was also during the 1970s that President Siaka Stevens and the All Peoples Congress (APC) government (1968–1985) gained control of the diamond sector, marking the beginning of a long decline for the industry. Diamonds quickly became a key strategic resource for Stevens' regime, as he appointed many of his allies to positions of power and rewarded them with diamond revenues. But these hegemonic strategies soon began to erode Stevens' hold on power, as official NDMC revenues plummeted and the diamond industry became a sphere of unregulated private enterprise. Official diamond exports fell from 1.7 million carats in the 1960s to a mere 50,000 carats by 1985 (Temple 2006). Government exchange and price controls and exchange rate overvaluation in the 1970s and 1980s produced a lucrative underground trade involving smuggling diamonds to purchase scarce imports for sale in black markets with the profits ploughed back into diamond smuggling. Those benefiting from the informal economy in Kono had access to subsidised imports of rice and consumer goods, but the majority was left with an invidious choice between 'tributing' and near-subsistence farming. Politicians taking a populist stance still managed to attract strong local support, but were frequently suppressed or co-opted by the ruling elite.

Under conditions of rising crime and civil unrest, state paramilitaries as well as private security forces began to be deployed to protect the mining interests of the politically well-connected against the 'illicit' mining activities of the rural poor and politically excluded.

Ultimately however, regime discipline began to erode under the weight of market informalisation and bureaucratic rent-seeking. The collapse began after Stevens retired in 1985 in favour of former army commander J.S. Momoh. It was accelerated further by a changing international political climate that saw increasingly stringent conditionalities attached to foreign loans. Government revenues plummeted from 17% of GDP in the 1970s to 8% in 1985–1989, yet a civil service bloated by patronage and subsidies on basic imports ensured that government expenditure remained high. By the mid-1980s, increasing scarcity of foreign currency and price controls had left basic imports of rice and fuel permanently in short supply. Infrastructure deteriorated, public sector real wages plummeted and government services gradually became non-functional. The Revolutionary United Front (RUF) insurgency of March 1991 served as the final catalyst for state collapse.

Civil war, displacement and post-conflict return, 1991–2011

The causes of Sierra Leone's decade-long conflict were multifaceted and complex, and much has been written about the political economy of the war and the role that diamonds played in the conflict (Keen 2005; Richards 2003). There is now a vast literature focusing on so-called 'blood diamonds', with particular attention revolving around the 'greed vs. grievance' debate (Collier 2000; Berdal and Malone 2001). While some observers have argued that the point of the war may not actually have been to win it, but rather 'to engage in profitable crime under the cover of warfare' (Smillie 2000, 24), others maintain that there is little evidence to suggest that diamonds were the fundamental cause of the conflict (Richards 2003). There does, however, appear to be some consensus that diamonds played a key role in fuelling and prolonging the war, as various parties to the conflict undoubtedly funded their war efforts through mining activities.

Of significance to the focus of this article is the impact that forced displacement from Kono District had during the civil war years, when rebel insurgents took control of Kono's diamond fields and local residents were forced to flee to the safety of Freetown. One of the demographic consequences of forced displacement during the civil war was that the country's population became much more mobile and more urban (Brown et al. 2005). It is estimated that more than 500,000 farming families were displaced by the conflict and agricultural production was so severely dislocated that by 2001 only 20% of the annual rice requirement (the staple food) was produced in the country (EIU 2002). According to the 2004 national census, the population of the Western Area almost doubled between 1985 and 2004 and youths were at the forefront of this influx.

Following the end of the war in 2002, many ex-combatants remained in Freetown, preferring to reside in a larger and more anonymous urban location. For this segment of the population, the prospect of returning to a rural setting to engage in farming activities has not been an attractive option, particularly since there is a risk of being exposed and held accountable for wartime atrocities committed (Binns and Maconachie 2005). As Peters (2007, 13) notes, 'towns, like the mining fields, offer the best social and economic niches for reintegration for those ex-combatants dubious about returning to home areas'. As such, in addition to those who remained in Freetown, what might be described as a 'second diamond rush' into Kono District took place as vast numbers of ex-combatants were drawn back to the mining settlements in search of diamonds. Many did not formally undergo the Disarmament, Demobilisation and Reintegration (DDR) process and the high concentration of former combatants in the area without skills and training was a source of great concern in Sierra Leonean government and donor circles. While security concerns during the colonial era centred on the fear that a flood of migrant youth to mining towns would serve to break the bonds of traditional communities, thereby undermining the system of 'indirect rule', more recent post-war concerns focus on the risk of a return to violence if the economic oppression and political exclusion of youth does not cease. Indeed, 15–30-year-old males are the very demographic group that is most likely to resume warfare if left frustrated and excluded (MSI 2004).

Present-day conditions in the growing towns in diamondiferous regions continue to provide ample grounds for grievance amongst miners and communities, and have fuelled tension between various actors. Although mining may represent an opportunity for some young people – both men and women – to escape the unequal circumstances of their agrarian communities, many have found themselves in equally unjust situations in mining areas. As is noted in a recent paper by Dale (2008, 17):

> Frequently...youth find the circumstances they were trying to escape in their hometowns replicated in their new locales, where they are strangers without strong social ties, living on borrowed land, and working in harsh conditions for limited incomes. The intersection of uneven income distributions, poor working conditions, large migrant or transient populations, access to drugs and alcohol, and illicit industries can create the potential for both small and large-scale conflict.

In the aftermath of the war, estimates suggest that the population of Koidu grew to 111,800 (Government of Sierra Leone 2006) and is now one of the most ethnically diverse cities in the country. As in the past, however, 'class' solidarity amongst young miners has overlapped with a politics of recognition in which ethnic Kono rights and entitlements become a banner for collective action and protest against external agencies perceived to be exploiting local resources unfairly. The proliferation of Kono-based mining advocacy groups in the post-war period has been striking. In the last couple of years especially, there has been a great rise in NGO and CSO activity, particularly concerned with mining issues. In some cases, protests have turned violent, such as occurred during the disturbances that took place at the Koidu Holdings blasting area in December 2007. However, on balance, the rise of civil society groups in Kono's mining towns has been a positive development.

Today, some of the worst poverty in Sierra Leone is concentrated in diamond mining towns. A 2005 Partnership Africa Canada study noted that the vast majority of artisanal miners live below the poverty line, often subsisting on less than one US dollar a day (PAC and Global Witness 2005). According to the Sierra Leonean government's 2005 Poverty Reduction Strategy Paper, Kono District – currently home to the largest concentration of artisanal miners in the country – has a far higher incidence of poverty than most surrounding districts where agriculture is the predominant livelihood activity. For example, the paper indicates that rural Kono has a poverty level of 79.6% compared to rural Pujehun's 59.6%, which is a mainly agricultural area. Alternatively, urban Kono has 56.3% of 'total poor' compared to Freetown's 17.1% (PAC 2006, 4). Youth are particularly vulnerable, and since many young miners are migrant 'strangers' who lack the social ties and family support to sustain them in host communities, coping with poverty is particularly difficult.

At the same time, however, current government estimates suggest that artisanal mining, Sierra Leone's second largest employer after agriculture, continues to provide a livelihood for some 120,000 to 200,000 people.[1] In 2007, US$141 million of diamonds were officially exported from the country, of which over US$100 million came from artisanal production (SKI/NMJD 2010; DDI 2008). Although, on the surface, artisanal mining activity in Sierra Leone appears to be completely anarchic, there is, in fact, a tightly managed, highly ordered structure to production, and such local-level mining remains vital to both local and national economies. It is significant to note that even though there are great challenges in directing diamond revenues through official channels, the artisanal sector does generate a considerable amount of revenue for the government, and at this micro-level the future potential of alluvial diamond mining in Sierra Leone is considerable (Ministry of Mineral Resources 2005).

While diamonds continue to be an important livelihood component for many living in Kono District, recent research carried out in the settlement of Kayima in 2004, and then again in 2008, is instructive in exploring how changing circumstances around alluvial mining activities may be altering livelihood patterns, and the relationships between mining towns and their surrounding rural areas. Kayima, now a settlement of 1881 (Government of Sierra Leone 2006), has long served as a readily available pool of mining labour, as it is located within easy reach of Koidu and other more locally situated alluvial mining areas such as Yengema (Maconachie and Binns 2007). Past stories of inter-generational conflict in Kayima are common, particularly over incidents that concern diamond mining. Yet more recent evidence from Kayima suggests that over the last five years developments may have taken place to diminish some of the reported tension between youth and chiefs, especially that driven by the control of traditional institutions which were used to coerce labour from young people and block their social advancement.

During the first stage of the research carried out in Kayima between May and July 2004, 50 households were randomly sampled and semi-structured interviews were conducted with a broad cross-section of society to explore questions around diamond mining, rural development and post-conflict reconstruction. Discussions specifically focused on issues concerning the relationships between youth, diamond mining and agriculture, and understanding the circular migration patterns that define the 'interlocking nature' of mining and farming (Maconachie and Binns 2007). During follow-up research carried out between September and November 2008, Kayima was revisited and another 50 households were randomly sampled and interviewed to see what had changed over the previous four years. While many of the same questions originally asked in 2004 were revisited, the interview schedule also revolved around a variety of other questions which explored the present relationship between youth and chiefs, changing attitudes towards artisanal mining, and associational life and club activity associated with farming.

As was the case in the research carried out in 2004, most of the households interviewed in 2008 reported that residents of Kayima still had strong links with nearby mining areas. When questioned about the present-day links that existed between the diamond mining and farming economies, interviewees believed that the most important relationship between the two concerned the profitable sale of foodstuffs to miners in nearby towns: 84% of the sample reported that farmers from Kayima obtained higher prices by taking their produce to mining areas to sell; 60% of respondents noted that mining supporters came to buy produce in Kayima; and 36% of the sample claimed that there was a strong nexus of seasonal work between farming and mining activities, whereby farmers went to mine in the dry season when there was less work to do on their plots.

However, in contrast, 92% of respondents interviewed in 2008 also reported that there were presently less youth in Kayima who were mining on either a full-time or seasonal basis. The 2008 data indicated that only 36% of the interview sample claimed that either they, or a member of their family, were engaged in diamond mining, whereas in 2004, 78% claimed that this was so. This decline in ASM participation is also reflected in official statistics from the Ministry of Mineral Resources and Political Affairs (MMRPA) for the period between 2007 and 2009, which illustrate a sharp decline in the total number of artisanal mining licenses granted in Kono and the country's four other major mining districts (Figure 2).

The reasons for the decline in artisanal mining were varied (Table 2), but according to interviewees, the overwhelming belief was that as alluvial gravels had become overworked and mined-out over the years, there were far fewer dividends in mining.[2] Many young miners interviewed claimed that they had spent years working as diamond diggers and had, in fact, seen very little in the way of any remuneration.[3] According to Temple (2006), the tendency for households in mining communities to diversify their livelihood portfolios, particularly when mining is not providing adequate dividends, indicates that mining is not merely a coping strategy but an accumulation strategy. She notes: 'people don't mine because they need to, [in order] to survive. They mine to be in a position to re-invest in other areas of their life such as business, education, housing and the various forms of savings' (Temple 2006, 24).

In response to a perceived decline in artisanal diamond mining activities and reports of lower prices being offered locally for diamonds,[4] interviewees reported that many young people in Kayima were now much more interested in rebuilding agricultural livelihoods. This appeared to be especially so in light of rising global

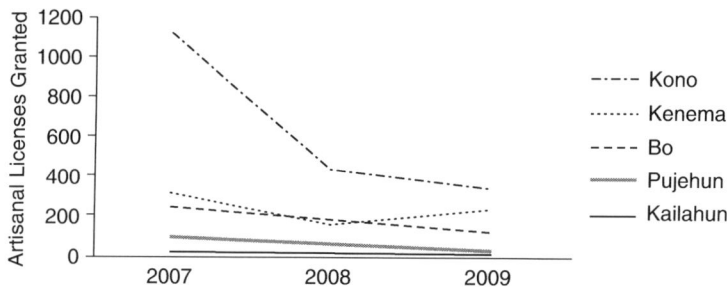

Figure 2. Mining licences issued between 2007 and 2009.
Source: Adapted from MMRPA (2007, 2008, 2009).

Table 2. Summary of reasons why fewer youth in Kayima are mining (2008).

Reason why fewer youth are mining (mentioned in interview)	% respondents who referred to reason
There are lower dividends in mining now	72
Youth realise that farming is worth investing in (more sustainable than mining)	24
The high price of rice (especially since 2007) has encouraged more youth to go into farming	12
Life is more expensive and diamond mining offers little return	12
Diamonds can now only be mined successfully by those individuals with access to large amounts of capital	12
Mining companies have stopped hiring	8
Small business activities and petty trading are more attractive than mining	8
Mining is a gamble	8
Elderly parents are in Kayima and they require full-time care	4

Source: authors' field data.

food costs since 2007 – many individuals believed that not only could they not afford to purchase rice, but if they could produce a surplus, it was now more lucrative to sell. The analysis suggests that mining households not only respond directly to global processes, but they also exercise trade-offs between productive activities to maximise their livelihood benefits. In this sense, it may be the case that a movement of young male migrants back to their villages of origin is not necessarily representative of a process of de-urbanisation, but is rather indicative of a process of re-agrarianisation. It would seem that, in this situation, the agricultural option extends beyond simply offering a vital livelihood fall-back for individuals. For some, farming activities and the re-agrarianisation of livelihoods may present a more lucrative and sustainable future than diamond mining can presently offer (Maconachie 2011). Indeed, evidence suggests that a number of entrepreneurial farmers in Kayima have generated a significant amount of income through the cultivation of cash crops, such as coffee, cocoa and citrus fruits, for sale to middlemen in the nearby town of Koidu (Maconachie and Binns 2007).

Elsewhere, in his recent study of the post-war reintegration process in the diamondiferous region of Tongo Field (Kenema District), Peters (2006) argued that some young men may now also be looking to agriculture as a means of escaping pre-war violence and labour exploitation associated with diamond mining. In the research carried out in Kayima, however, the motivation to move into agriculture appeared to be purely economic, as many young men saw better opportunities for livelihood advancement in farming, as they had given up on the hope of one day finding a big diamond. A number of NGOs have realised the importance of encouraging young diamond miners to return to their villages, by providing them with the resources necessary to re-engage in farming activities. In some cases, NGOs have also had to act as mediators, as many youth have feared retribution from community elders for wartime atrocities that they had committed.[5]

Many households in 2008 also reported that a large number of people in Kayima were now trying to plant larger farms, and it was suggested that this was increasingly possible since family farming labour was more plentiful (due to the increased number of people that had returned to Kayima following the war, and the perception that there were now fewer youth who were leaving to go to the diamond mining areas). Research carried out in Kayima in 2004 suggested that a major bottleneck in rejuvenating agriculture immediately after the war was that many farmers did not possess the means to activate traditional labour cycles (Maconachie 2008). However, 98% of the sample interviewed in 2008 believed that reciprocal relationships were now much stronger, especially those to do with labour. Among the many changes that have taken place, it would appear that one significant difference over the four-year period is that extreme poverty levels have been reduced, meaning that more resources may now be available (labour and monetary surplus), and so engaging in reciprocal farming arrangements (*boma*) may be easier. This has been an important factor in convincing many young miners that farming may now be a much more lucrative and worthwhile livelihood activity.

Conclusion

In providing an historical overview of mining expansion and contraction over time, this article has explored the significant impacts that mining has had on the rural–urban continuum in Sierra Leone, and how this has shaped political, economic and social change in diamondiferous regions. Above all, the analysis demonstrates that diamonds have had diverse and varying impacts on urban populations over time, in many cases having a transformative effect on the physical development of Sierra Leone's towns and cities. In some instances, diamonds and the seasonal migration of labour to mining towns have provided great impetus for local economic development through various regional multiplier effects, driving demand for agricultural, industrial and service sector activities. Yet at other points in history, diamonds have been the source of significant social and political upheaval.

While the impact of mining in Sierra Leone has thus long been a mixed blessing for those living and working in diamondiferous communities, there continues to be much debate as to the role that diamonds might assume in shaping both regional urban development, and the country's future development trajectory more broadly. On one hand, urban areas in Kono District continue to attract young mining migrants in search of a better life. But on the other hand, as alluvial diamond gravels have increasingly become worked-out in recent years, the declining demand for unskilled labour in the mining industry may emerge as one of the greatest challenges for overcoming the social and economic marginalisation of youth in mining areas and related security problems. However, on a positive note, evidence also suggests that the decline of artisanal mining and the rise of large-scale industrial mining have reawakened the interest of many young ex-miners in farming, especially those who enjoy hereditary land rights. While one consequence of the war may be that the population is now more urban and more mobile, the return of young people to their villages of origin, and rapprochement with local chiefs, may ultimately be helping to drive a resurgence of community-based cooperation in rural Kono District, a development which could provide a more durable basis for sustainable and democratic development in the years to come.

Acknowledgements

This paper draws on historical analysis previously developed in a co-authored paper written with Richard Fanthorpe entitled 'Mining for change: Diamonds, "youth" politics and emerging public spaces in post-conflict Sierra Leone', presented at the 'Mining across generations: Artisanal and small-scale mining workshop', 17 January 2009, Cambridge, UK. The author is grateful to Dr Fanthorpe for his insights and contribution to the analysis. The fieldwork upon which the paper is based was funded by a small grant from the Nuffield Foundation.

Notes

1. Personal communication with the Director of the Ministry of Mineral Resources and Political Affairs (MMRPA), Freetown, 22 November 2010.
2. This sentiment was evident in the typical response of interviewees, 'diamonds don loss', a Krio saying which literally translated means 'diamonds are much fewer'.
3. Diamond diggers who work at the bottom of the supply chain must endure particularly dangerous and unhealthy conditions, with most being paid two cups of rice and less than a dollar a day. In 2008, 67% of those from Kayima who reported that they were miners claimed that they had never earned a significant amount of money from their activities. Approximately 33% of miners said they had only earned very small amounts of money over the years (typically between 50,000 and 100,000 Leones).
4. Reports indicate that the global market for wholesale diamonds in 2009 shrank from US$21.5 billion to US$12 billion, with the price of polished gems dropping by an average of 30% from its peak in August 2008.
5. Interview with the Director of GTZ Kono, Koidu, 25 September 2008.

Note on contributor

Roy Maconachie is Lecturer in International Development at the Centre for Development Studies at the University of Bath. A development geographer by training, his research is concerned with the political economy of natural resource management and the socio-economic dimensions of small-scale mining in West Africa. His recent work in Sierra Leone has been actively engaged with debates around the relationship between small-scale mining and agrarian change, in particular exploring the links between diamond mining, livelihoods and post-conflict development. He has published on the topic in the *Journal of International Development, Natural Resources Forum, Journal of Development Studies* and *African Affairs* amongst others. He can be contacted at: R.Maconachie@bath.ac.uk

References

Amco-Robertson Mineral Services Ltd. 2002. *Sierra Leone diamond policy study.* Freetown: Department for International Development.
Berdal, M., and D. Malone, eds. 2001. *Greed and grievance: Economic agendas in civil wars.* Boulder, CO: Lynne Rienner.
Binns, J.A. 1982. The changing impact of diamond mining in Sierra Leone. *Research Papers in Geography* no. 9, University of Sussex, Brighton.
Binns, J.A., and R. Maconachie. 2005. 'Going home' in post-conflict Sierra Leone: Diamonds, agriculture and re-building of rural livelihoods in the Eastern Province. *Geography* 90, no. 1: 67–78.
Brown, T., R. Fanthorpe, J. Gardener, L. Gberie, and M. Gibril Sesay. 2005. *Sierra Leone: Drivers of change.* Bristol: The IDL Group.
Collier, P. 2000. *Economic causes of civil conflict and their implications for policy.* Washington, DC: The World Bank.
Dale, P. 2008. *Access to justice in Sierra Leone: A review of the literature.* Justice for the Poor Publication, May 2008. Washington, DC: The World Bank.

Diamond Development Initiative (DDI). 2008. *Standards and guidelines for Sierra Leone's artisanal diamond mining sector.* Ottawa, Canada: Diamond Development Initiative.

Economist Intelligence Unit (EIU). 2002. *Country profile: Guinea, Sierra Leone, Liberia.* London: EIU.

Gamble, D.P. 1964. Kenema – a growing town in Mende Country. *Bulletin of the Sierra Leone Geographical Association* 7, no. 8: 9–12.

Gberie, L. 2005. *A dirty war in West Africa: The RUF and the destruction of Sierra Leone.* London: Hurst and Co.

Government of Sierra Leone. 1954. *Annual Report on the Administration of the Provinces for the Year 1953.* Freetown: Government Printer.

Government of Sierra Leone. 1956. *1955 Report on the Administration of the Provinces.* Freetown: Government Printer.

Government of Sierra Leone. 2006. *Population and housing census – 2004. Analytical report on population distribution, migration and urbanization in Sierra Leone.* Freetown: Government Printer, November.

Hayward, F. 1972. The development of a radical political organization in the bush: A case study of Sierra Leone. *Canadian Journal of African Studies/Revue Canadienne des Études Africaines* 6, no. 1: 1–28.

Keen, D. 2005. *Conflict and collusion in Sierra Leone.* Oxford: James Currey.

Maconachie, R. 2008. New agricultural frontiers in post-conflict Sierra Leone? Exploring institutional challenges for wetland management in the Eastern Province. *Journal of Modern African Studies* 46, no. 2: 235–66.

Maconachie, R. 2011. Re-agrarianizing livelihoods in post-conflict Sierra Leone? Mineral wealth and rural change in artisanal and small-scale mining communities. *Journal of International Development* 23: 1054–67.

Maconachie, R., and T. Binns. 2007. 'Farming miners' or 'mining farmers'? Diamond mining and rural development in post-conflict Sierra Leone. *Journal of Rural Studies* 23: 367–80.

Management Systems International (MSI). 2004. *Integrated diamond management in Sierra Leone: a two-year pilot project.* Report prepared for assistance of the United States Agency for International Development, Washington, DC.

Ministry of Mineral Resources. 2005. *Details of policy measures relating to small scale and artisanal mining and marketing of precious minerals.* Freetown: Mines Division, Ministry of Mineral Resources.

Ministry of Mineral Resources and Political Affairs (MMRPA). 2007. *DACDF payments and artisanal mining licences 2007.* Freetown: MMRPA.

Ministry of Mineral Resources and Political Affairs (MMRPA). 2008. *DACDF Payments and artisanal mining licences 2008.* Freetown: MMRPA.

Ministry of Mineral Resources and Political Affairs (MMRPA). 2009. *DACDF payments and artisanal mining licences 2009.* Freetown: MMRPA.

Partnership Africa Canada. 2004. *Diamond Industry Annual Review, Sierra Leone 2004.* Freetown, Sierra Leone: The Diamonds and Human Security Project, Partnership Africa Canada.

Partnership Africa Canada (PAC). 2006. *Diamond industry annual review, Sierra Leone 2006.* Freetown: The Diamonds and Human Security Project, Partnership Africa, February.

Partnership Africa Canada (PAC) and Global Witness. 2005. *Rich man, poor man – development diamonds and poverty diamonds: The potential for change in the alluvial diamond fields of Africa.* Ottawa: Partnership Africa Canada.

Peters, K. 2006. Footpaths to reintegration: Armed conflict, youth and the rural crisis in Sierra Leone. PhD diss., University of Wageningen.

Peters, K. 2007. From weapons to wheels: Young Sierra Leonean ex-combatants become motorbike taxi-riders. *Journal of Peace Conflict and Development* 10 March 2007. http://www.peacestudiesjournal.org.uk (accessed May 28, 2012).

Reno, W. 1995. *Corruption and state politics in Sierra Leone.* Cambridge: Cambridge University Press.

Richards, P. 1996. *Fighting for the rainforest: War, youth and resources in Sierra Leone.* Oxford: James Currey.

Richards, P. 2003. *The political economy of internal conflict in Sierra Leone.* Working Paper 21. Netherlands Institute of International Relations. Clingendael: Conflict Research Unit, August 2003.

Riddell, J.B. 1974. Periodic markets in Sierra Leone. *Annals of the Association of American Geographers* 64, no. 4: 541–8.

Rosen, D. 1974. The Kono of Sierra Leone – A study of social change. Mimeo paper to Kono Road Project. Freetown: Fourah Bay College.

Saylor, R.G. 1967. *The economic system of Sierra Leone.* Durham, NC: Duke University Press.

Smillie, I. 2000. Getting to the heart of the matter: Sierra Leone, diamonds and human security. *Social Justice* 27, no. 4: 24–31.

Street Kids International and Network Movement for Justice and Development (SKI/NMJD). 2010. *One day I will do something else: Realizing the potential of Sierra Leonean youth.* Street Kids International, Network Movement for Justice and Development and Diamond Development Initiative, February.

Swindell, K. 1974. Sierra Leonean mining migrants: Their composition and origins. *Transactions of the Institute of British Geographers* 61: 47–64.

Temple, H. 2006. Livelihoods report. Unpublished DfID Report.

United Kingdom Government, Sierra Leone. 1956. *Annual Report of the Colonial Office.* Freetown, Sierra Leone: Government Printer.

Van der Laan, H.L. 1965. *The Sierra Leone diamonds.* Oxford: Oxford University Press.

Zack-Williams, A. 1995. *Tributors, supporters and merchant capital: Mining and underdevelopment in Sierra Leone.* London: Edwin Mellen Press.

Index